Cognitive Radio Networks

ANALOG CIRCUITS AND SIGNAL PROCESSING

Series Editors:
Mohammed Ismail, The Ohio State University
Mohamad Sawan, École Polytechnique de Montréal

For further volumes:
http://www.springer.com/series/7381

Ahmed Khattab • Dmitri Perkins
Magdy Bayoumi

Cognitive Radio Networks

From Theory to Practice

Springer

Ahmed Khattab
Department of Electronics and Electrical Communications Engineering
Cairo University
Giza, Egypt

Dmitri Perkins
The Center for Advanced Computer Studies
University of Louisiana at Lafayette
Lafayette, LA, USA

Magdy Bayoumi
The Center for Advanced Computer Studies
University of Louisiana at Lafayette
Lafayette, LA, USA

ISBN 978-1-4614-4032-1 ISBN 978-1-4614-4033-8 (eBook)
DOI 10.1007/978-1-4614-4033-8
Springer New York Heidelberg Dordrecht London

Library of Congress Control Number: 2012944181

Printed on acid-free paper

Springer is part of Springer Science+Business Media (www.springer.com)

To my great parents, Ahmed Khattab
To my beloved family, Dmitri Perkins
To my dear students, Magdy Bayoumi

Preface

Cognitive Radio Networks (CRNs) are the key enabling technology for realizing Opportunistic Spectrum Access (OSA) to alleviate the severe spectrum underutilization and provide a solution to the wireless spectrum scarcity. OSA refers to the communications paradigm in which the communicating parties dynamically exploit the spectrum bands that are not utilized by the primary wireless services licensed to operate over such bands. CRNs are foreseen as the future of wireless communication technologies that provide wireless connectivity for emerging services. The main component of CRNs, and hence OSA, is the cognitive radio transceiver. A cognitive radio is a wireless device that senses the surrounding radio environment and opportunistically accesses the unutilized spectrum band(s) based on its assessment of the activities of the surrounding primary licensed networks. OSA in distributed ad-hoc CRNs is the focus of significant research interest, especially from a theoretical perspective. The resulting theoretical OSA approaches are challenged by the practical limitations of existing cognitive radios. The focus of this book is on the less well-studied issue of implementing distributed Opportunistic Spectrum Access given practical radio transceiver technologies.

This book distinguishes itself from the existing affluent literature of CRNs and OSA. Existing literature can be classified into two categories: One that presents a self-contained introduction of the emerging Cognitive Radio Networking paradigm outlining the theoretical fundamentals and requirements for enabling such a technology. The emphasis of such books is on the theoretical design, optimization, and performance evaluation of Opportunistic Spectrum Access in CRNs. The second—and more related—category is mainly concerned with the implementation of CRNs using software-defined radios (SDR). While SDR provided seamless flexibility in the design and implementation of CRNs due to the fact that the design is carried out in software, SDR-based CRN implementations lag orders of magnitude behind realistic communication speeds. In contrast, this book:

- Incorporates the practical implementation constraints and issues in the theoretical formulation of the Opportunistic Spectrum Access problem in distributed CRNs

and presents a first-of-its-kind analytically optimized framework for designing distributed ad-hoc CRNs based on contemporary radio transceiver technologies.
- Provides an in-depth illustration of the implementation of such a framework that shows that Opportunistic Spectrum Access and CRNs can indeed be implemented using traditional hardware rather than only being restricted to the SDR implementations presented in the related literature that cannot achieve the high communication speeds of today's world.
- Quantifies the gains that existing theoretical approaches can attain by adopting the individual components of the proposed practical OSA framework that were designed in lieu of the common practices of contemporary radio technologies.

This book is the first to combine CRN theory with practice and target real-life technologies. It serves as a guide in the transition from conventional communications to cognitive radio communications without assuming currently unavailable transceiver technologies as the case with existing theoretical approaches which are only implementable using SDR platforms.

The main challenge of existing distributed opportunistic spectrum management schemes is that they do not consider the unavoidable practical limitations of today's Cognitive Radio Networks such as the inability to measure the interference at the primary receivers. Consequently, optimizing the constrained Cognitive Radio Network performance based only on the local interference measurements at the cognitive radio senders does not lead to truely optimal performance due to the existence of hidden or exposed primary senders. More specifically, existing schemes have a cognitive radio sender decide its transmission strategy based on its local interference measurement—while such decisions should have been made based on the interference measurement at the nearby primary receivers to be interfered with its transmission. However, there does not exist a practical mechanism that enables a cognitive radio to determine the interference at nearby primary receivers. Furthermore, existing transceiver technologies and spectrum measurement techniques are incapable of accurately assessing the spectrum usage over a wide frequency range due to the limitations imposed by the transceiver hardware.

Cognitive Radio Networks: From Theory to Practice presents a probabilistic framework for opportunistic spectrum management that optimizes the constrained cognitive user goodput while taking the unavoidable sources of spectrum sensing inaccuracy into account. This book introduces the Rate-Adaptive Probabilistic (RAP) spectrum management approach that (1) probabilistically explores individual spectrum bands as local interference measurements lead to inaccurate spectrum access decisions and (2) adopts a non-greedy probabilistic spectrum access policy that prevents a single cognitive transmission from monopolizing an available spectral opportunity, and hence allows multiple cognitive flows to fairly share the available bandwidth without explicit coordination. In contrast, existing opportunistic spectrum management techniques greedily use the highest possible power/rate only over the best frequency band thought to be available. Furthermore, such techniques require explicit coordination between different cognitive flows to ensure fairness in opportunity sharing.

The book presents the analytical formulation of the CRN performance optimization problem of the proposed RAP approach as a mixed integer non-linear program to derive its optimal probabilities and powers/rates. Packet-level simulations are used to evaluate the RAP performance and demonstrate its gains in arbitrary large-scale ad-hoc networks. The large-scale simulation results show that such a probabilistic approach achieves up to 138% higher goodput compared to greedy and hypothetically optimal approaches with significantly better fairness at the expense of slightly higher primary network outages within their permissible limits.

Finally, the proposed Opportunistic Spectrum Access approach is implemented using the Wireless open Access Research Platform (WARP) to demonstrate its superior performance in a real system. Our empirical results show that the proposed RAP approach indeed achieves multiple folds increase in the achieved goodput (up to 673% gain) compared to existing Opportunistic Spectrum Access approaches especially with highly active primary networks. We conclude that despite relying on simple cognitive radio hardware, this book presents a practical distributed opportunistic spectrum management approach which outperforms existing theoretical approaches that assume fully capable cognitive radios.

This book is an extension of the Ph.D. dissertation of Dr. Ahmed Khattab submitted to the University of Louisiana at Lafayette under the supervision of Dr. Dmitri Perkins and Dr. Magdy Bayoumi. We started this work at the Center of Advanced Computer Studies (CACS) at the University of Louisiana at Lafayette in 2009 using the facilities of the Wireless Systems and Performance Engineering Research (WiSPER) and the Very Large-Scale Integration (VLSI) research laboratories.

Cognitive Radio Networks: From Theory to Practice targets a wide range of readers including but not limited to researchers, industry experts, and graduate students. On the one hand, readers with theoretical interests will experience an unprecedented treatment of the conventional CRN performance optimization problem that takes into account the practical limitations of today's technologies. On the other hand, readers interested in real-life distributed CRN realization will be exposed to a first-of-its-kind clean-slate implementation approach that demonstrates the significant multi-faced performance improvement achieved by practical OSA.

This book offers the reader a range of interesting topics portraying the current state-of-the-art in cognitive radio technologies. In simple terms, while several existing Opportunistic Spectrum Access approaches have been developed and theoretically optimized, they are challenged by the inherent constraints of practical implementation technologies. Analyzing these constraints and proposing an attractive and practical solution to counter these limitations are the basic aims of this book. This book presents the set of practical OSA system components that the interested reader will be able to glean information not only to incorporate into his/her own particular OSA design problem, but also most of all to experience an enjoyable and relatively effortless reading, providing the reader with intellectual stimulation.

Lafayette, LA

Ahmed Khattab
Dmitri Perkins
Magdy Bayoumi

Acknowledgements

The authors would like to acknowledge the support of the National Science Foundation (NSF). This work was partially supported under the NSF Career Grant No. 0448055. We would also like to acknowledge the U.S. Department of Energy (DOE) due to their partial support of this work under Award Number DE-FG02-04ER46136. The role of the State of Louisiana, Louisiana Board of Regents is also acknowledged under Contract Numbers DOE/LEQSF(2004-07)-ULL and LEQSF(2003-06)-RD-A-35. We are also thankful to Dr. Nian-Feng Tzeng and Dr. Hong-Yi Wu for their precious comments and feedback that helped us further improve the material presented in this book. The authors would like to warmly thank the Springer editorial team for their support and assistance.

Dr. Khattab would like to dedicate a special expression of gratitude and appreciation to his family, especially his beloved parents: Khattab and Sanaa, for their patience and full emotional support. He owes it all to them.

Dr. Perkins would like to thank his students for their dedication and constant stream of probing questions. He would also like to say a very special thank you to his wife, Coretta, and four children—Paige, Naomi, Lillian, and Mason for their laughter, love, and support.

Dr. Bayoumi would like to thank his students, former and current, for enriching his life and keeping him young in heart and spirit. They are making the academic life exciting, interesting, and never boring.

Contents

List of Figures

List of Tables

Acronyms

ACK	ACKnowledgement
AWGN	Additive White Gaussian Noise
BPSK	Binary Phase Shift Keying
BS	Base Station
CORVUS	COgnitive Radio approach for usage of Virtual Unlicensed Spectrum
CPE	Consumer Premise Equipment
CPU	Central Processing Unit
CRN	Cognitive Radio Network
CROSS	Cognitive Radio Open Source System
CSMA/CA	Carrier Sense Multiple Access with Collision Avoidance
DARPA	Defense Advance Research Projects Agency
DRiVE	European Dynamic Radio for IP services in Vehicular Environment
DSA	Dynamic Spectrum Access
DySPAN	Dynamic SPectrum Access Networks
EIRP	Effective Isotropically Radiated Power
FCC	Federal Communications Commission
FPGA	Field Programmable Gate Array
FSU	Flexible Spectrum Use
GNU	GNU's Not Unix (recursive acronym)
IEEE	Institute of Electrical and Electronics Engineers
ISM	Industrial, Scientific and Medical
JFI	Jain's Fairness Index
MAC	Medium Access Control
MIMO	Multiple-Input Multiple-Output
OFDM	Orthogonal Frequency Division Multiplexing
OPT-MAC	OPTimal Medium Access Control
OSA	Opportunistic Spectrum Access
OverDRiVE	Spectrum Efficient Uni- and Multi-cast Services Over Dynamic Radio Network in Vehicular Environments
PHY	PHYsical
PRN	Primary Radio Network or Primary Network

PU	Primary User
PU-RX	Primary Receiver
PU-TX	Primary Transmitter
QAM	Quadrature Amplitude Modulation
QPSK	Quadrature Phase Shift Keying
RAM	Random Access Memory
RAP	Rate-Adaptive Probabilistic
RAP-MAC	Rate-Adaptive Probabilistic Medium Access Control
RF	Radio Frequency
RSSI	Received Signal Strength Indicator
RTS/CTS	Request-To-Send/Clear-To-Send
SCC41	Standardization Coordinating Committee 41
SDR	Software-Defined Radio
SG	Spectrum Grant
SR	Spectrum Request
SU	Secondary User
SU-RX	Secondary Receiver
SU-TX	Secondary Transmitter
TV	Television
UDP	User Datagram Protocol
UHF	Ultra High Frequency
USB	Universal Serial Bus
USRP	Universal Software Radio Peripheral
VHF	Very High Frequency
VLSI	Very Large-Scale Integration
WARP	Wireless open-Access Research Platform
WiNC2R	Network Centric Cognitive Radio Platform
WiSPER	Wireless Systems and Performance Engineering Research
WLAN	Wireless Local Area Network
WNaN	Wireless Network after Next
WRAN	Wireless Regional Area Networks
XG	neXt Generation

Chapter 1
Introduction

The wireless communications technology has recently extended beyond personal communication services to safety monitoring, wearable/embedded health monitoring, disaster relief, traffic control, and broadband access at high mobility. Such proliferation of wireless communications is leading today's world to a paradigm shift towards a seamless and ubiquitous wireless connectivity where end users or devices are always connected to a vast amount of services. The exponentially growing demand for wireless connectivity is challenged by the scarcity of the radio spectrum since much of the usable spectrum has already been allocated to certain services.

Over the past few decades, the wireless spectrum has been treated as a static quantity to be divided into well-defined blocks. The current allocation policy of the wireless spectrum is to statically grant exclusive rights to licensees or services on a long-term basis over vast geographical areas [1]. This allocation is governed by auction processes conducted by regulatory authorities, such as the Federal Communications Commission (FCC) in the United States and the corresponding organizations in other countries, that grant a given spectrum block to the highest bidder. The main mission of such regulatory authorities is to assure the orderly use of the spectrum and gain the greatest return on this natural resource for the citizens of any given nation. Therefore, the target of the spectrum auctions is to sell the spectrum to the highest bidders to raise the governmental revenues. Consequently, this makes the efficient use of the wireless spectrum a secondary goal for such organizations.

Fortunately, the regulatory organizations dedicate a certain amount of bandwidth to be freely available for low power operation. Such unlicensed spectrum bands, e.g., the 900 MHz, 2.4 GHz and 5 GHz industrial, scientific and medical (ISM) bands, have become very cowered. The transmission over such unlicensed bands is governed by different standards such as the Institute of Electrical and Electronics Engineers (IEEE) 802.15.4 and IEEE 802.11 standards. However, the interference resulting from the coexistence of a vast number of geographically-collocated networks, either for similar or different wireless services, limits the bandwidth available for new services. Even though several wireless communication schemes,

A. Khattab et al., *Cognitive Radio Networks: From Theory to Practice*,
Analog Circuits and Signal Processing, DOI 10.1007/978-1-4614-4033-8_1,

such as the Multiple-Input Multiple-Output (MIMO) technology, have recently emerged to achieve high data rates while being resilient to interference, such schemes are not necessarily applicable to all wireless applications. For example, miniature wearable and embedded sensors used for health care applications cannot have multiple antennas due to area constraints.

Recently, the FCC [2, 3], and other organizations such as the Defense Advance Research Projects Agency (DARPA) [4, 5], have shown that large portions of the licensed spectrum bands are vastly underutilized. More specifically, the measurements in [2] showed that the licensed spectrum is underutilized for 15–85% of the time depending on the spatial location. Other measurement studies conducted by academic and/or industrial organizations have shown similar trends. For instance, Fig. 1.1 depicts the results of a measurement study of the spectrum utilization in some rural areas that show that the average spectrum utilization over a 3-day period of the frequency bands below 3 GHz is only 13.1% and 17.5% in New York City and Chicago, respectively [6]. Such measurements illustrate the severe inefficiency of the utilization of the wireless spectrum resulting from the static long-term allocation policy.

The significant underutilization of the licensed spectrum coupled with the heavy overutilization of the unlicensed portion of the wireless spectrum has motivated the need for a new spectrum management paradigm. The new spectrum management paradigm aims at providing a solution to the wireless spectrum scarcity problem that tackles new wireless services while improving the efficiency of the utilization of the licensed spectrum bands. Such paradigm allows the wireless terminals to exploit the temporal and spatial variations of the utilization of different spectrum bands as shown in Fig. 1.2. Hence, it provides a communication capacity for future wireless services. This spectrum management paradigm is referred to as Opportunistic Spectrum Access (OSA), Dynamic Spectrum Access (DSA), or Flexible Spectrum Use (FSU). In this book, we shall use the term Opportunistic Spectrum Access to refer to such a communications paradigm.

1.1 Opportunistic Spectrum Access in Cognitive Radio Networks

Opportunistic Spectrum Access (OSA) refers to the communications paradigm in which the communicating parties *dynamically* exploit the spectrum bands that are not utilized by the primary wireless services licensed to operate over such bands. OSA must not disturb the performance of a licensed service by immediately vacating its spectrum band once a licensed user is willing to use its legitimate spectrum. In which case, OSA implies that the communicating parties will search for another unutilized spectrum band to exploit during the absence of activities of its licensed users. Figure 1.3 conceptually illustrates Opportunistic Spectrum Access. The lack of licensed users' activities is called a spectral opportunity, a spectral hole, or a white space. For the rest of the book, we shall use the term spectral opportunity.

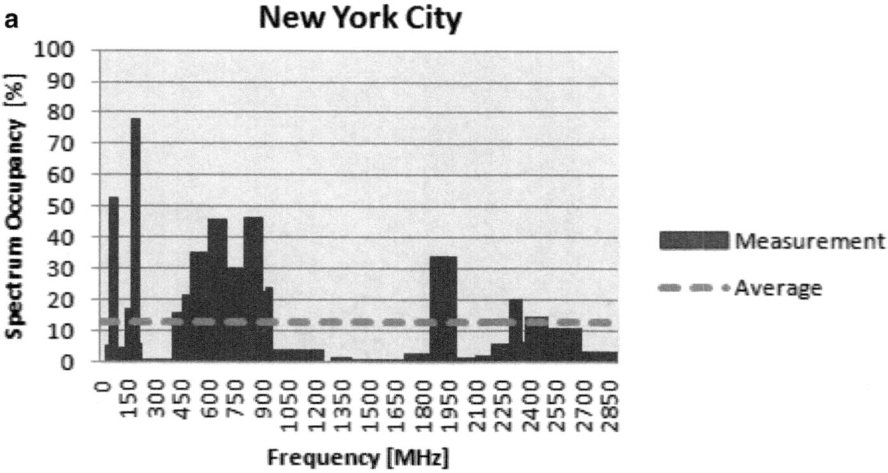

New York city measurements show that the average spectrum utilization is only 13.1%.

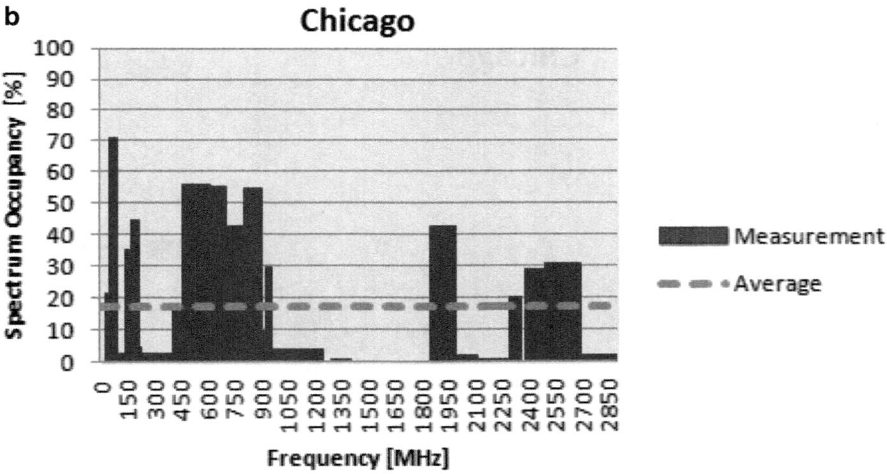

Chicago measurements show that the average spectrum utilization is only 17.5%.

Fig. 1.1 Measured spectrum occupancy averaged over a 3-day period in the New York City and Chicago areas. Figure is created using the data reported in [6]

As shown in Fig. 1.3, OSA is to dynamically track spectral opportunities to provide wireless access.

The Cognitive Radio Network (CRN) is the key enabling technology for implementing Opportunistic Spectrum Access to alleviate the severe spectrum underutilization and provide a solution for spectrum scarcity [7, 8]. CRNs are foreseen as the future of wireless communication technologies that provide wireless connectivity for emerging services. The key enabling component of CRNs, and

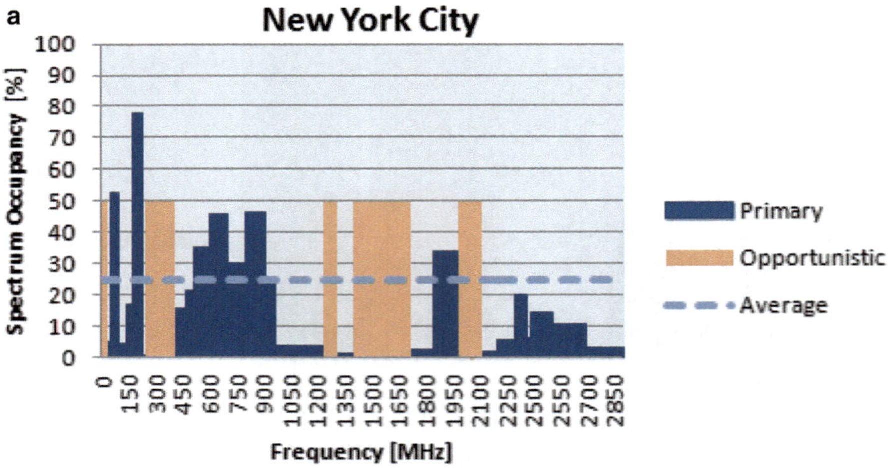

Opportunistic spectrum access can improve the average spectrum occupancy by 90% in New York city even if the unutilized spectral opportunities are only used for 50% of the time.

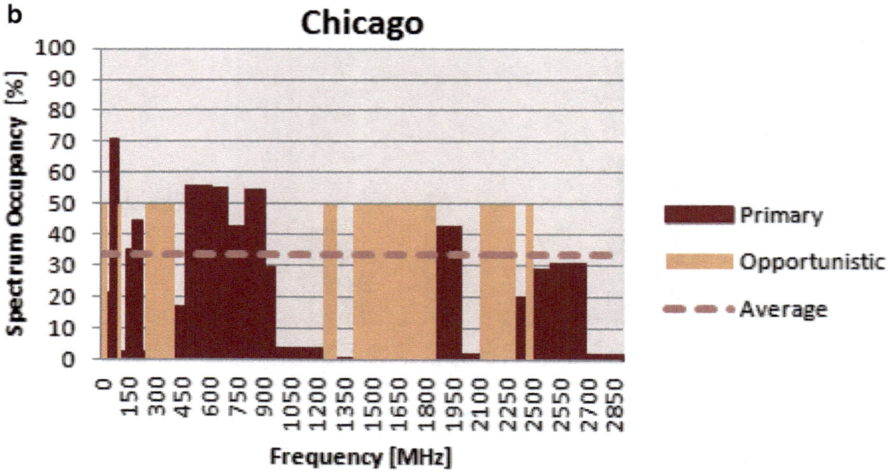

Opportunistic spectrum access can improve the average spectrum occupancy by 93.3% in Chicago even if the unutilized spectral opportunities are only used for 50% of the time.

Fig. 1.2 Opportunistic Spectrum Access significantly improves the spectrum utilization by exploiting otherwise unutilized frequency bands without interfering with the licensed spectrum owners. The primary utilization pattern is generated using the data reported in [6]

hence Opportunistic Spectrum Access, is the cognitive radio transceiver. A cognitive radio is a wireless device that senses the surrounding radio environment and opportunistically accesses the unutilized spectrum band(s) based on its assessment of the activities of the surrounding primary licensed networks [9, 10].

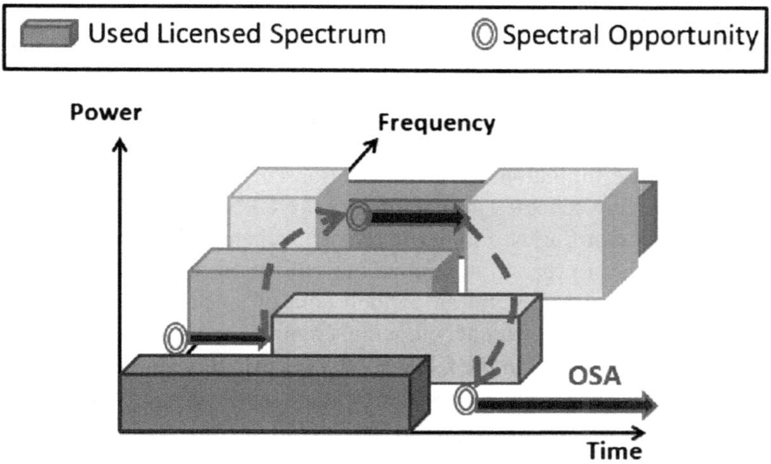

Fig. 1.3 Opportunistic Spectrum Access tracks the available spectral opportunities

Cognitive Radio Networking and Opportunistic Spectrum Access are challenged by several factors that make such a networking scheme fundamentally different from traditional wireless communications schemes. On one hand, OSA requires the wireless terminals to be able to identify the available spectral opportunities and take decisions regrading the most appropriate spectrum band to use. This poses many stringent requirements on the transceiver hardware implementation. For instance, a cognitive radio transceiver should be able to sense the radio activities over a wide range of the spectrum (e.g., multiple gigahertz) and process the acquired measurements to come up with access decisions in very short time periods. While significant efforts are currently being made to provide such high performance transceivers at low cost, existing transceiver technologies still cannot allow using Cognitive Radio Networking at its full potential.

On the other hand, the existence of the primary licensed networks in the radio environment of a CRN makes Opportunistic Spectrum Access fundamentally different from traditional medium access. OSA wireless access necessitates the ability of the wireless terminals to not disturb the transmissions of the primary users. Furthermore, OSA must turn the spectrum access rights back to the licensed owners once they are active. Furthermore, cognitive radio users within a CRN should be able to coordinate their transmissions in order to share the available spectral opportunities. Such characteristics of opportunistic access are not required for conventional wireless access in which the wireless terminals operate on one or more predefined set of frequency bands and legitimately use such band(s) without the need to abandon their access rights. Hence, conventional medium access mechanisms developed for traditional wireless networks are not suitable for use in Opportunistic Spectrum Access networks.

Several opportunistic spectrum sensing and management schemes have been proposed in the literature aiming at optimizing the CRN goodput while not degrading the performance of licensed networks [11–24]. However, such schemes do not take into account the practical limitations of CRNs. More Specifically, the CRN users are not able to measure the interference at the primary receivers, and instead, take their spectrum access decisions based on their local measurements of the interference due to the primary senders. Such inaccurate sensing, combined with the speed and other hardware limitations of cognitive transceivers, does not guarantee the optimality of existing schemes in maximizing the CRN goodput or satisfying the primary networks constraints. Furthermore, such schemes adopt greedy spectrum access mechanisms in order to maximize the utilization of any available spectral opportunities, and hence, require an explicit control mechanism for coordinating the spectrum decisions of different competing cognitive transmissions. Such greedy strategies lead to unfairness in the CRN goodput distribution and cause the control mechanism to be a bottleneck and a single point of failure of the system [7, 8].

1.2 Book Scope and Objectives

This book explores Cognitive Radio Networking from a practical point of view. While the CRN technology is a significant enabling technology for increasing the spectrum efficiency, building a fully-functional wireless CRN presents numerous challenges. The IEEE 802.22 standard is the first standard for CRNs in which a set of base stations uses spectrum databases to provide wireless access to the customer-premise equipments (CPEs). In contrast, this book targets a distributed CRN model in which cognitive radio nodes individually make their spectrum access decisions. We aim at taking the practical limitations of cognitive radio transceivers into account while modeling the CRN behavior in such a network architecture. Furthermore, we target having the different cognitive flows fairly sharing the available spectral opportunities without having individual sender-receiver pairs explicitly coordinating their decisions with each other (i.e., no explicit inter-flow coordination). Thereby, we reduce the overall system complexity (e.g., by omitting the need for a network-wide synchronization mechanism) and avoid the overhead associated with explicit inter-flow coordination which negatively impacts the CRN goodput. This contrasts with existing distributed CRN approaches which adopt one form or another of inter-flow coordination [18–24].

The book objective is to design, optimize, and realize a practical spectrum management scheme for distributed CRNs that (1) counters the *unavoidable* inaccuracies in spectrum measurements and their consequent negative impact on the performance of both the CRN and the primary licensed networks, and (2) allows cognitive radio users to fairly share the spectral opportunities without explicit inter-flow coordination. The goal of the proposed scheme is to relax the hardware requirements of the cognitive transceivers. Our main idea is to use randomized sensing combined with a probabilistic and non-greedy transmission policy to counter the inaccuracies in spectrum sensing and prevent an individual cognitive radio flow from exclusively exploiting an available opportunity. We address the following two open questions. First, given that a cognitive radio sender does not apriori know the impact of its transmission on nearby primary receivers, how aggressive/conservative a cognitive radio sender should/should not be to alleviate spectral miss-predictions and missed opportunities. Second, how non-greedy spectrum access can allow multiple cognitive radio users to share spectral opportunities without explicit information sharing.

1.3 Book Contributions

In this book, we consider a distributed asynchronous CRN wherein cognitive nodes are not necessarily within each other's range. The inherent inaccuracies in spectrum sensing of cognitive radios are unavoidable due to not only hidden and exposed primary transmitters but also hidden and exposed cognitive radio terminals. We present a practical spectrum management framework to counter such inaccuracies and minimize their impact on the performance of both the primary and Cognitive Radio Networks.[1] The book contributions are as follows.

First, we propose the Rate-Adaptive Probabilistic (RAP) spectrum management framework and its Medium Access Control protocol realization (RAP-MAC). The main ideas behind our framework are: (1) Any spectrum band can be explored with a certain probability—even if the measured interference level is high—since the local interference measurements at the CRN senders do not infer the interference at nearby primary receivers. (2) A CRN transmission does not greedily exploit a spectral opportunity. Instead, a CRN transmission probabilistically switches between the maximum permissible transmission power/rate and lower powers/rates. Thereby, RAP-MAC probabilistically reduces the potential harm to nearby primary receivers and leaves a spectral margin for other CRN flows to transmit. In a distributed multiuser environment, RAP-MAC adaptively makes different CRN flows share the spectral opportunities without explicit inter-flow coordination. In contrast, hypothetically-optimal spectrum management schemes greedily transmit only over the band(s) with the least primary interference at the maximum permissible power/rate relying on an explicit interflow coordination mechanism.

Second, we analytically formulate the constrained CRN optimization problem according to the RAP-MAC framework in order to compute the optimal probabilities of transmission and the used powers and rates. In our formulation, we consider another practical limitation of CRN hardware that is only a finite set of transmission powers/rates is available. This limitation causes our optimization problem to be a mixed-integer non-linear program which complexity is NP-complete. We present an exhaustive study of the impact of various factors on the optimal RAP-MAC parameter values. More specifically, we investigate the impact of the outage constraints and the activity pattern of the primary networks on the optimal probabilities of the RAP-MAC protocol as well as the achievable cognitive user goodput.

[1] A preliminary version of the framework was presented in [25].

Next, we use packet-level simulations to demonstrate that RAP-MAC probabilistic spectrum management achieves up to 138% higher goodput compared to greedy spectrum management depending on the CRN traffic demand. This superior performance is attributed to the RAP-MAC probabilistic sensing and transmission policies which explore more spectral opportunities and lead to fewer transmission failures compared to deterministic and hypothetically-optimal spectrum management. Furthermore, RAP-MAC results in different CRN flows fairly sharing the available opportunities without explicit inter-flow coordination. Meanwhile, greedy spectrum management results in 47% of the flows receiving less than 10% of the average goodput. Our approach satisfies the primary network performance constraints despite the use of cognitive transceivers with narrow-band sensing capability compared to hypothetically-optimal spectrum management that assumes wide-band cognitive transceivers. However, our approach results in slightly higher outages to the primary networks. The superior goodput performance of the proposed RAP approach outweighs the excess primary networks' outages.

Finally, we use the state-of-the-art hardware platform to demonstrate the performance gains of the proposed RAP-MAC protocol in a real system. We develop a custom cross-layer framework for implementing the opportunistic spectrum management mechanisms that are common to a wide range of OSA protocols. We use this framework to implement the proposed RAP-MAC protocol as well as a suite of other OSA protocols that represent different spectrum sensing and spectrum access approaches. We perform an extensive set of experiments that do not only demonstrate the RAP-MAC performance gains but also explain how much is the contribution of each of its components in the overall gain. Our experimental results show that RAP-MAC achieves approximately 70% goodput gain at low primary network activities compared to the protocols that adopt deterministic and greedy access mechanisms based on wide-band sensing (i.e., the hypothetically-optimal protocols). The RAP-MAC gain increases to be multiple folds as the primary networks carry more traffic (e.g., 194% and 673% gains at 75% and 100% primary networks' activity factors, respectively). Our experiments show that 70–80% of the RAP-MAC is attributed to its random sensing approach and the remaining 20–30% is due to its non-greedy probabilistic access mechanism. We show that such individual gains can be attained by legacy OSA protocols by adopting either of the two main components of RAP-MAC. The reported experimental results emphasize the superior RAP-MAC goodput performance that comes at the expense of slightly higher primary network outages within the maximum permissible limits.[2]

[2]The basics of the hardware implementation framework were introduced in [26] and a preliminary set of the experimental results is presented in [27].

1.4 Book Organization

The remainder of the book is organized as follows. In Chap. 2, we present the necessary preliminaries of cognitive radios and opportunistic access networks. We briefly highlight the design guidelines of CRNs and their impact on different Opportunistic Spectrum Access functions. We review the state-of-the-art of Cognitive Radio Networking and Opportunistic Spectrum Access in Chap. 3. In Chap. 4, we define the system model considered in the remainder of the book and motivate our goal in such a network model. We propose the Rate-Adaptive Probabilistic (RAP) framework for Opportunistic Spectrum Access and its protocol implementation RAP-MAC in Chap. 5. Next, we formulate the constrained CRN performance optimization problem in terms of the RAP-MAC parameters in order to compute their optimal values and investigate their impact on the overall performance in Chap. 6. In Chap. 7, we exhaustively study the performance of RAP-MAC in an arbitrary large-scale network setup via simulations. We then present our implementation of the proposed protocol using the state-of-the-art hardware platform in Chap. 8. Using this implementation we demonstrate the RAP-MAC superior performance in a real system and quantify the gains of its individual components in Chap. 9. Finally, we conclude the book and propose future research directions in Chap. 10.

References

1. FCC: ET Docket No 02-135 Spectrum policy task force report. Federal Communications Comission (FCC) (2002)
2. FCC: ET Docket No 02-155 Spectrum policy task force report. Federal Communications Comission (FCC) (2002)
3. FCC: ET Docket No 03-222 Notice of proposed rule making and order. Federal Communications Comission (FCC) (2003)
4. Darpa XG working group: The XG architectural framework rfc v1.0. http://www.darpa.mil/ato/programs/XG/rfc_af.pdf (2003). Last Accessed 30 October 2011
5. Darpa XG working group: The XG vision rfc v1.0. http://www.darpa.mil/ato/programs/XG/rfc_vision.pdf (2003). Last Accessed 30 October 2011
6. McHenry, M.A., Tenhula, P.A., McCloskey, D., Roberson, D.A., Hoo, C.S.: Chicago spectrum occupancy measurements & analysis and a long-term studies proposal. http://www.wtapas.org/final-papers/ChicagoSpectrum-McHenry-Session-I-1.pdf (2005). Accessed 25 July 2012
7. Akyildiz, I.F., Lee, W.Y., Chowdhury, K.R.: CRAHNs: Cognitive radio ad hoc networks. Ad Hoc Networks (Elsevier) **7**(5), 810–836 (2009)
8. Salameh, H.B., Krunz, M.: Channel access protocols for multihop opportunistic networks: Challenges and recent developments. IEEE Networks **23**(4), 14–19 (2009)
9. Mitola III, J.: Cognitive radio: An integrated agent architecture for software defined radio. Ph.D. thesis, KTH Royal Institute of Technology (2000)
10. Mitola III, J.: Cognitive radio for flexible mobile multimedia communication. In: Proceedings of IEEE International Workshop on Mbile Multimedia Communications (MoMuC), San Diego, CA (1999)
11. Yucek, T., Arslan, H.: A survey of spectrum sensing algorithms for cognitive radio applications. IEEE Comm. Surv. Tutorials **11**(1), 116–130 (2009)

12. Tian, Z., Giannakis, G.: Compressed sensing for wideband cognitive radios. In: Proceedings of IEEE ICASSP, Honolulu, HI (2007)
13. Liu, H., Krishnamachari, B.: Randomized strategies for multi-user multi-channel opportunity sensing. In: Proceedings of IEEE CCNC Cognitive Radio Networks Workshop, Las Vegas, NV (2008)
14. Liang, Z., Liu, W., Zhou, P., Gao, F.: Randomized multi-user strategy for spectrum sharing in opportunistic spectrum access network. In: Proceedings of IEEE ICC Workshops, Beijing, China (2008)
15. Raman, C., Yates, R.D., Mandayam, N.B.: Scheduling variable rate links via a spectrum server. In: Proceedings of IEEE DySPAN 2005, Baltimore, MD (2005)
16. Lotfinezhad, M., Liang, B., Sousa, E.S.: Optimal control of constrained cognitive radio networks with dynamic population size. In: Proceedings of IEEE INFOCOM 2010, San Diego, CA (2010)
17. Hosseinabadi, G., Manshaei, M.H., Hubaux, J.P.: Spectrum sharing games of infrastructure-based cognitive radio networks. Tech. rep. http://infoscience.epfl.ch/record/128112?ln=en (2008)
18. Zhao, Q., Tong, L., Swami, A., Chen, Y.: Decentralized cognitive MAC for opportunistic spectrum access in ad hoc networks: a POMPD framework. IEEE J. Sel. Area. Comm. **25**(3), 589–600 (2007)
19. Huang, S., Liu, X., Ding, Z.: Opportunistic spectrum access in cognitive radio networks. In: Proceedings of IEEE INFOCOM 2008, Phoenix, AZ (2008)
20. Wang, F., Krunz, M., Cui, S.: Price-based spectrum management in cognitive radio networks. IEEE J. Sel. Top. Signal Process. **2**(1), 74–87 (2008)
21. Xu, H., Li, B.: Efficient resource allocation with flexible channel cooperation in OFDMA cognitive radio networks. In: Proceedings of IEEE INFOCOM 2010, San Diego, CA (2010)
22. Salameh, H.B., Krunz, M., Younis, O.: MAC protocol for opportunistic cognitive radio networks with soft guarantees. IEEE Trans. Mobile Comput. **8**(10), 1339–1352 (2009)
23. Anandkumar, A., Michael, N., Tang, A.: Opportunistic spectrum access with multiple users: Learning under competition. In: Proceedings of IEEE INFOCOM 2010, San Deigo, CA (2010)
24. Chaporkar, P., Proutiere, A., Asnani, H.: Learning to optimally exploit multi-channel diversity in wireless systems. In: Proceedings of IEEE INFOCOM 2010, San Diego, CA (2010)
25. Khattab, A., Perkins, D., Bayoumi, M.A.: Probabilistic framework for opportunistic spectrum management in cognitive ad-hoc networks. EURASIP J. Wireless Comm. Networking **2011**(188) (2011)
26. Khattab, A., Perkins, D., Bayoumi, M.A.: Opportunistic Spectrum Access: From Theory to Practice. IEEE Vehicular Technology Magazine. **7**(2), 62–68 (2012)
27. Khattab, A., Perkins, D., Bayoumi, M.A.: Design, Implementation and Characterization of a Practical Distributed Cognitive Radio Network, submitted to IEEE Transactions of Wireless Communications (2012)

Chapter 2
Cognitive Radio Networking Preliminaries

2.1 Cognitive Radio Technology

A cognitive radio is the key technology that allows a cognitive wireless terminal to dynamically access the available spectral opportunities. A cognitive radio was defined by Mitola in his seminal work as "a radio or system that senses, and is aware of, its operational environment and can dynamically and autonomously adjust its radio operating parameters accordingly" [1, 2]. This definition was generalized by the FCC to be "a radio or system that sense its electromagnetic environment and can dynamically and autonomously adjust its radio operating parameters to modify system operation, such as maximize throughput, mitigate interference, facilitate interoperability, access secondary markets" [3]. From these definitions, a cognitive radio has two key features that distinguish it from a traditional radio: the cognition capability and the reconfigurability. Figure 2.1 illustrates how these unique features of a cognitive radio conceptually interact with the radio environment. This illustration is referred to as the cognition cycle that is continually run by the cognitive radio to observe spectral opportunities, create plans to adapt itself, decide, and act to explore the best opportunities.

2.1.1 Cognition Capability of a Cognitive Radio

The cognition capability of a cognitive radio is defined as the ability of the cognitive radio transceiver to sense the surrounding radio environment, analyze the captured information and accordingly decide the best course of action(s) in terms of which spectrum band(s) to be used and the best transmission strategy to be adopted. Such a cognition capability allows a cognitive radio to continually observe the dynamically changing surrounding radio environment in order to interactively come up with the appropriate transmission plans to be used. The three main components of the cognitive radio cognition cycle can be briefly explained as follows.

A. Khattab et al., *Cognitive Radio Networks: From Theory to Practice*,
Analog Circuits and Signal Processing, DOI 10.1007/978-1-4614-4033-8_2,

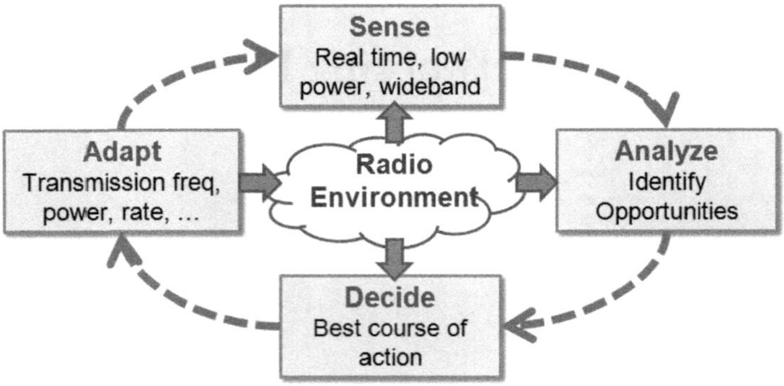

Fig. 2.1 Functional architecture of a cognitive radio

2.1.1.1 Spectrum Sensing

Spectrum sensing refers to the ability of a cognitive radio to measure the electromagnetic activities due to the ongoing radio transmissions over different spectrum bands and to capture the parameters related to such bands (e.g., cumulative power levels, user activities, etc.). Spectrum sensing is one of the most critical functions of a cognitive radio as it provides the awareness of the spectrum usage in the surrounding environment [4]. A cognitive radio must make real-time decisions about which bands to sense, when, and for how long. The sensed spectrum information must be sufficient enough for the cognitive radio to reach accurate conclusions regarding the radio environment. Furthermore, spectrum sensing must be fast in order to track the temporal variations of the radio environment. Such requirements of spectrum sensing puts stringent requirements on the hardware implementation of cognitive radios in terms of the sensing bandwidth, the processing power, the radio frequency (RF) circuitry, etc. Existing spectrum sensing techniques depend on detecting the activities of the primary transmitters. Such schemes are generally classified to matched filter detection, energy detection, feature detection, and interference temperature measurement.

2.1.1.2 Spectrum Analysis

Spectrum analysis is to infer the existence of spectral opportunities in the surrounding radio environment based on the sensed radio environment parameters. A spectral opportunity is conventionally defined as "a band of frequencies that are not being used by the primary user of that band at a particular time in a particular geographic area" [5]. However, such a definition is not general enough as it covers only three dimension of the spectrum space: frequency, time, and space. Other dimensions of a given spectrum can be exploited. For example, the coding dimension which utilizes spread spectrum coding techniques to create spectral opportunities over a given spectrum band currently utilized by its licensed users. Similarly, the angle dimension creates spectral opportunities through the use of beamforming to allow the cognitive radio users to simultaneously transmit over a currently utilized band. Furthermore, the recent advancements in radio transmission techniques, such as the use of Multiple-Input Multiple-Output (MIMO) at the physical layer, present new dimensions in the definition of a spectral opportunity. For instance, stream control and antenna selection [6] can be used to allow cognitive radio users to simultaneously transmit with the primary licensed users without degrading the performance of such legitimate users. Due to the existence of different dimensions of a spectrum, we use the following generalized definition a spectral opportunity in the remainder of the book: A spectral opportunity is "a theoretical hyperspace occupied by radio signals, which has dimensions of location, angle of arrival, frequency, time, and possible others" [4, 7, 8].

2.1.1.3 Spectrum Access Decisions

The last step of the cognition cycle of a cognitive radio is to decide the set of transmission actions to be taken based on the outcome of the spectrum sensing and analysis procedures. More specifically, a cognitive radio utilizes the information gathered regarding the spectrum bands identified as available spectral opportunities to define the radio transceiver parameters for the upcoming transmission(s) over such frequency bands. The set of transceiver parameters to be decided depends on the underlying transceiver architecture. Examples of the action set can include which spectrum is more favorable for an upcoming transmission, the time instant a transmission over a certain band should start, the maximum transmission power, the modulation rate, the spread spectrum hopping scheme, the angle of arrival for directional transmissions, and the number and identity of the antennas to be used in MIMO systems, etc. Based on the sensed spectrum information and the transceiver architecture, a cognitive radio defines the values of the parameters to be configured for an upcoming transmission.

2.1.2 Reconfigurability of a Cognitive Radio

The second key feature that distinguishes a cognitive radio from a traditional one, and completes the cognition cycle depicted in Fig. 2.1, is its ability to re-tune its transceiver parameters on the fly based on its assessment of the surrounding radio environment. While today's radios have considerable flexibility in terms of their ability to reconfigure some transmission parameters such as the transmission rate and power, they are typically designed to operate over certain frequency band(s) according to a certain communication protocol. A cognitive radio transceiver should be more flexible than just this in order to be able to exploit emerging spectral opportunities over a wider spectrum range. For instance, a cognitive radio must be able to configure the transmission bandwidth to adapt to spectral opportunities of different sizes. Furthermore, a cognitive radio cannot be constrained to a certain communication protocol. Instead, a cognitive radio must determine the appropriate communication protocol to be used over different spectral opportunities based its recognition of the radio environment.

In his seminal work, Mitola devised the software-defined radio as the ideal implementation environment of radios with seamless configuration capabilities (since parameter configuration is performed using software commands). Thus, cognitive radios were originally referred to as software radios with extended self-awareness capability [2]. However, software-defined radios cannot fulfill the data rate requirements of most of today's wireless services due to the software/hardware interface bottleneck [9, 10]. Thus motivated, a significant research effort has been—and is currently being—made towards realizing a fast multi-gigahertz cognitive radio transceiver hardware with the seamless configuration flexibility of software-defined radios at low cost.

2.1.2.1 Spectrum Mobility

The reconfigurability of a cognitive radio transceiver reflects the spectrum mobility function introduced by Mitola in his definition of the cognitive radio [1, 2]. Spectrum mobility refers to the process in which a cognitive radio terminal changes its frequency of operation. In order to maintain seamless wireless connectivity, a cognitive radio terminal should be able to switch to a new frequency band upon either the appearance of the primary licensed user(s) of the current band or the deterioration of the channel quality of the currently used channel. In other words, spectrum mobility is the cognitive radio functionality that actually allows the cognitive radio to dynamically explore the available spectral opportunities. Thus, spectrum mobility is associated with a handoff mechanism that guarantees the transition to the new frequency band without breaking (or significantly degrading the quality of) the communication between communicating cognitive radio terminals. While the cognition functions of the cognitive radio mainly affect the lower layers of the CRN, namely the physical and medium access layers, spectrum mobility and handoff also affect higher layers. Hence, spectrum mobility schemes should ensure smooth and fast frequency transition and protocol/paramter adjustment in order to minimize the latency that could harm the performance of higher layer protocols. Even though mobility-based handoff mechanisms have been investigated in the context of cellular networks and can be used to lay the foundation for spectrum mobility, CRN spectrum mobility poses several new challenges. However, spectrum mobility is beyond the scope of this book.

2.2 Cognitive Radio Network Architectures

A typical CRN environment consists of a number of Primary Radio Networks (PRNs) that coexist within the same geographical area of a single CRN (also referred to as the secondary network). A primary network is an existing network that is licensed to operate in a certain spectrum band. Hence, a primary network is also referred to as a licensed network. Primary networks can either be based on a centralized infrastructure or distributed ad-hoc in nature. The users of a primary network can only access the spectrum licensed to this particular network. Primary users have priority with respect to spectrum access and operate as they are the sole users of their licensed spectrum. Hence, primary users do not provide any type of cooperation with the secondary network. PRNs are non-intrusive and the transmissions of the primary users should not be affected by the secondary users. Therefore, the primary networks define upper bounds on the CRN activities in their licensed bands, typically in terms of maximum power levels, to guarantee the promised performance level to their legitimate users.

On the other hand, the CRN is not licensed to operate in a predefined band. Spectrum access for the CRN is achieved in an opportunistic manner that allows the secondary users to opportunistically access the entire spectrum available to *all* of the geographically-collocated PRNs. Recall that the cognitive users can also exploit the unlicensed spectrum. This is referred to as spectrum heterogeneity of CRNs [11, 12]. When operating in a licensed band, the CRN transmissions must adhere to the constraints imposed by its primary owner. A CRN can either be centralized infrastructure-based network or a distributed ad-hoc network as shown in Fig. 2.2.

2.2.1 Centralized Cognitive Radio Networks

Centralized CRNs are infrastructure-based networks in which cognitive radio base stations control and coordinate the transmission activities of the secondary cognitive radio users as shown in Fig. 2.2a. The cognitive radio base stations control the secondary transmissions over both the licensed and unlicensed bands by collecting all the spectrum-related information from the cognitive radio users. Based on the collected information, the base stations take global spectrum access decisions for all nodes. An example centralized infrastructure-based CRN is the IEEE 802.22 network model. The IEEE 802.22 is the first world-wide standard for CRNs [13]. The IEEE 802.22 standard defines the specifications of a point-to-multipoint communication scheme over the unused television (TV) bands in which a base station manages cognitive radio users within 33 km radius using a centralized spectrum database. Other examples include the European Dynamic Radio for IP services in Vehicular Environment (DRiVE) [14] and Spectrum Efficient Uni- and Multi-cast Services Over Dynamic Radio Network in Vehicular Environments

Fig. 2.2 Cognitive Radio Network architectures

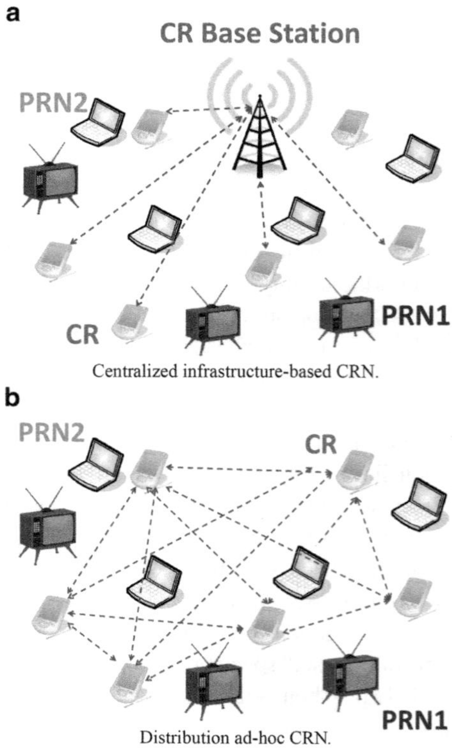

(OverDRiVE) [15] projects. These projects have a centralized entity that coordinates the dynamic utilization of the temporal and spatial spectral opportunities. Centralized infrastructure-based CRNs are beyond the scope of this book.

2.2.2 Distributed Cognitive Radio Networks

Alternatively, CRNs can also have the cognitive nodes communicating with each other via ad-hoc point-to-point connections over either the licensed or the unlicensed bands as shown in Fig. 2.2b. While alleviating the infrastructure cost, such infrastructureless CRNs have increased networking complexity. In the absence of a controlling centralized entity, cognitive radio nodes in a distributed CRN jointly coordinate their spectrum access decisions to share the available spectral opportunities. Thus, global mechanisms such as network-wide synchronization might be needed for spectrum access coordination. In addition, distributed cooperative detection and communication techniques are used to improve the overall network performance. Example distributed CRNs include, the peer-to-peer mode of DARPA's neXt Generation (XG) dynamic access network [16, 17], DARPA's Wireless Network after Next (WNaN) military testbed [18], the Nautilus distributed scalable and efficient coordination project for open spectrum ad-hoc network [19, 20], and the cognitive radio approach for usage of virtual unlicensed bands (CORVUS) [21]. This book targets Opportunistic Spectrum Access in distributed CRNs. Our goal is to alleviate the network-wide coordination overhead by omitting inter-flow communications in such a network model.

2.3 Guidelines of Cognitive Radio Networking

The coexistence of the primary networks within the CRN environment distinguishes CRNs from other traditional networks. The CRN transmissions should not disturb the transmissions within the primary networks. This constraint does not exist in legacy wireless networks. Therefore, Medium Access Control (MAC) protocols developed for such networks (more specifically, those developed for multi-channel and/or multi-radio networks) are not well suited to the unique characteristics of CRNs. The existence of the primary users makes the Opportunistic Spectrum Access problem fundamentally different from the medium access problem in multi-channel networks since the latter problem is simply a resource sharing problem for users within a given network. In order to realize an Opportunistic Spectrum Access network, the following design guidelines are mandated [11, 12].

- An Opportunistic Spectrum Access network should be transparent to the users of the primary networks. Hence, no coordination is required between the primary and the secondary users.
- An Opportunistic Spectrum Access network should provide guarantees to the performance of the primary licensed networks.
- Cognitive radio nodes should make efficient and accurate spectrum sensing and spectrum access decisions while exploring either the unutilized or the utilized bands. These decisions should account for the dynamics of the time-varying activities of the primary users.
- The CRN should define a coordination mechanism (either explicit or implicit) to maximize the spectrum utilization efficiency and allow cognitive radio users to fairly share the available spectral opportunities.

2.4 Cognitive Radio Network Applications

Cognitive Radio Networking and Opportunistic Spectrum Access can be used in different applications. In what follows, we briefly discuss those applications that can benefit from the research conducted in this book.

2.4.1 Cognitive Mesh Networks

Multi-hop wireless mesh networks have recently gained significant popularity as a cost-effective solution for last-mile Internet access. Traditional wireless mesh network are challenged by the scarcity of the wireless bandwidth needed to meet the high-speed requirements of existing wireless applications. Opportunistic Spectrum Access can be used to alleviate the bandwidth scarcity problem of mesh networks by allowing the mesh nodes to dynamically explore any available spectral opportunities. Such cognitive mesh networks are meant be used to provide broadband access to rural, tribal, and other under-resourced regions [22].

2.4.2 Public Safety Networks

Public safety networks are another type of networks that can exploit Cognitive Radio Networking. Public safety networks are used for communications among police officers and fire and paramedic personnel. Such networks are also challenged by the limited amount of allocated spectrum. Even with the recent extensions of the allocated public safety spectrum bands, the public safety personnel do not have the technology to dynamically operate across the different spectrum segments. Recall that public safety licensees have a wide variety of bands available (VHF-Low, VHF-Hi, 220 MHz, UHF below 800, UHF-800, etc.). The cognitive radio technology can offer public safety networks more bandwidth through Opportunistic Spectrum Access. Furthermore, a public safety CRN can provide a substantial communication improvement by allowing the interpretability across different public safety services while smartly adapting to the high peak-to-average nature of the traffic carried out by such networks [23].

2.4.3 Disaster Relief and Emergency Networks

Natural disasters such as hurricanes, earthquakes, wild fires, or other unpredictable phenomena usually cause the communications infrastructure to collapse. For example, some base stations of cellular networks can fall, the connectivity between sensor nodes and the sink node in static wireless sensor networks can be lost, existing Wireless Local Area Networks (WLANs) can be damaged, etc. This results in a set of partially or fully damaged coexistent networks that were previously deployed and then became disconnected. Meanwhile, there is an urgent need for a means of communications to help the rescue teams to facilitate organized help, rehabilitation efforts, and to locate the disaster survivors. CRNs can be used for such emergency networks (e.g., see [24] and references therein). The use of Opportunistic Spectrum Access in disaster relief networks can provide a significant amount of bandwidth that can handle the expected huge amount of voice, video, and other critical and time-sensitive traffic. It is worth mentioning that WLANs were used in the relief of the Haiti earthquake. However, the communication over such a network was unreliable and suffered significant delays [25].

2.4.4 *Battlefield Military Networks*

Unfortunately, the recent advances in wireless technologies made the job of communication jamming and/or hacking much easier. Consequently, achieving reliable and secure communications in modern battlefields has become a more challenging task. Recall that a battlefield communication network provides the only means of communications between soldiers, armed vehicles, and other units in the battlefield amongst themselves as well as with the headquarters. This implies that such networks do not only require significant amount of bandwidth, but also mandate secure and reliable communications to carry vital information. The cognitive radio is the key enabling technology for realizing such densely deployed networks which use distributed Opportunistic Spectrum Access strategies to fulfill the bandwidth and reliability needs. Note that, the dynamic nature of OSA makes the ability to track and jam a communication more difficult. Thus motivated, DARPA initiated the Wireless Network after Next (WNaN) program aiming at creating a flexible architecture for military communications [18]. The main goal of the WNaN program is to develop a low-cost handheld cognitive radio terminal that is capable of selecting its own frequencies and forming a dense network within a large battlefield area.

2.4.5 *Leased Networks*

All of the aforementioned CRN applications have the secondary users exploiting the resources of the primary networks without being beneficial to the primary networks in any way. However, a primary network can benefit from leasing a fraction of its licensed spectrum to secondary operators adopting cognitive radio technology to opportunistically access the spectrum. The entrance of the secondary operator to the market of the incumbent primary network can increase the revenue of the primary licensed operator [26].

References

1. Mitola III, J.: Cognitive radio: An integrated agent architecture for software defined radio. Ph.D. thesis, KTH Royal Institute of Technology (2000)
2. Mitola III, J.: Cognitive radio for flexible mobile multimedia communication. In: Proceedings of IEEE International Workshop on Mbile Multimedia Communications (MoMuC), San Diego, CA (1999)
3. FCC: ET Docket No 03-108 Notice of proposed rule making and order: Facilitating opportunities for flexible, efficient, and reliable spectrum use employing cognitive radio technologies. Federal Communications Comission (FCC) (2005)

4. Yucek, T., Arslan, H.: A survey of spectrum sensing algorithms for cognitive radio applications. IEEE Comm. Surv. Tutorials **11**(1), 116–130 (2009)
5. kolodzy, P., et al.: Next generation communications: Kickoff meeting. In: Proceedings of the Defense Advanced Research Projects Agency (DARPA'01) (2001)
6. Gaur, S., Jiang, J.S., Ingram, M., Demirkol, M.: Interfering MIMO links with stream control and optimal antenna selection. In: Proceedings of IEEE Globecom'04, Dallas, TX (2004)
7. Matheson, R.: The electrospace model as a frequency management tool. In: Proceedings of International Symposium on Advanced Radio Technologies'04, Boulder, CO (2003)
8. Drozd, A.L., Kasperovich, I.P., Carroll, C.E., Blackburn, A.C.: Computational electromagnetics applied to analyzing the efficient utilization of the RF transmission hyperspace. In: Proceedings of IEEE/ACES International Conference on Wireless Communications and Applied Computational Electromagnetics, Honolulu, HI (2005)
9. Ng, M.C., Fleming, K.E., Vutukuru, M., Gross, S., Arvind, H.B.: Airblue: A system for cross-layer wireless protocol development. In: Proceedings of ACM/IEEE ANCS '10, Brooklyn, NY (2010)
10. Nychis, G., Hottelier, T., Yang, Z., Seshan, S., Steenkiste, P.: Enabling MAC protocol implementations on software-defined radios. In: Proceedings of USENIX symposium on NSDI, Boston, MA (2009)
11. Akyildiz, I.F., Lee, W.Y., Chowdhury, K.R.: CRAHNs: Cognitive radio ad hoc networks. Ad Hoc Networks (Elsevier) **7**(5), 810–836 (2009)
12. Salameh, H.B., Krunz, M.: Channel access protocols for multihop opportunistic networks: Challenges and recent developments. IEEE Networks **23**(4), 14–19 (2009)
13. IEEE Working Group on Wireless Regional Area Networks: Enabling rural broadband wireless access using cognitive radio technology in TV whitespaces. http://www.ieee802.org/22/. Accessed 25 July 2012
14. Xu, L., Tnjes, R., Paila, T., Hansmann, W., Frank, M., Albrecht, M.: DRiVEing to the internet: Dynamic radio for ip services in vehicular environments. In: Proceedings of the 25th Annual Conference on Local Computer Networks, LCN'00, Tampa, FL (2000)
15. Tnjes, R., Moessner, K., Lohmar, T., Wolf, M.: OverDRiVE spectrum efficient multicast services to vehicles. In: Proceedings of IST Mobile and Telecommunicatiosn Summit, Thessaloniki, Greece (2002)
16. Darpa XG working group: The XG architectural framework rfc v1.0. http://www.darpa.mil/ato/programs/XG/rfc_af.pdf (2003). Last Accessed 30 October 2011
17. Darpa XG working group: The XG vision rfc v1.0. http://www.darpa.mil/ato/programs/XG/rfc_vision.pdf (2003). Last Accessed 30 October 2011
18. DARPA's Wieless Network after Next Project: http://www.darpa.mil/Our_Work/STO/Programs/Wireless_Network_after_Next_(WNAN).aspx. Accessed 25 July 2012
19. Zheng, H., Cao, L.: Device-centric spectrum management. In: Proceedings of IEEE DySPAN 2005, Baltimore, MD (2005)
20. Zheng, H., Peng, C.: Collaboration and fairness in opprtunistic spectrum access. In: Proceedings of IEEE ICC 2010, Seoul, Korea (2005)
21. Brodersen, R., Wolisz, A., Cabric, D., Mishra, S., Willkomm, D.: CORVUS: A cognitive radio approach for usage of virtual unlicensed spectrum. Berkeley Wireless Research Center (BWRC) White paper (2004)
22. Steenkiste, P., Sicker, D., Minden, G., Raychaudhuri, D.: Future directions in cognitive radio network research. NSF Workshop Report (2009)
23. Gorcin, A., Arslan, H.: Public safety and emergency case communications: Opportunities from the aspect of cognitive radio. In: Proceedings of IEEE DySPAN 2008, Chicago, IL (2008)
24. Rehmani, M.H., Viana, A.C., Khalife, H., Fdida, S.: A cognitive radio based internet access framework for disaster response network deployment. Research Report RR-7285, INRIA. http://hal.inria.fr/inria-00482593/en/ (2010). Accessed 25 July 2012
25. Goldstein, H.: Engineers help NGOs get online after Haiti quake. IEEE Spectrum (2010)
26. Guijarro, L., Pla, V., Vidal, J.R.: Competition in cognitive radio networks: Spectrum leasing and innovation. In: Proceedings of IEEE CCNC 2011, Las Vegas, NV (2011)

Chapter 3
State-of-the-Art of Cognitive Radio Networks

3.1 Opportunistic Spectrum Sensing

The problem of finding which frequency bands to sense and probe before transmission has been widely addressed in the context of both multi-channel networks and CRNs (see [1] and references therein for a more detailed discussion). Recently, the focus of the related literature was to relax the assumptions/requirements of the sensing module of a cognitive radio. For instance, adopting only a subset of the available frequency bands to probe has been proposed in [2, 3] based on distributed learning techniques. In [4], the authors compute the network capacity when only a subset of the available frequency bands is to be used due to transceiver hardware constraints. Both adjacent and random channel assignment models were considered. Alternatively, relaxing the amount of information needed to assess the existence of spectral opportunities was addressed in [5–8]. Compressed sensing [5] techniques and randomized sensing [6, 8] and sampling [7] were proposed. However, all the aforementioned sensing techniques lead to inaccurate decisions in some scenarios (such as the hidden or exposed primary sender scenarios) since they only target the detection of the RF activities of the primary senders. Consequently, optimizing the CRN performance while overlooking such inherent inaccuracies does not lead to optimal performance in all scenarios.

One way to address the spectrum sensing limitations is to exploit the bidirectional communication nature in some primary networks [9, 10]. By monitoring the reverse traffic originating from the primary receivers, the secondary cognitive radio senders can infer the existence or the absence of nearby primary receivers. Another potential approach to collect information about nearby primary receivers is to detect the local oscillator leakage power emitted by the RF front-end of the primary receivers [11]. Since such leakage power is typically weak, implementing a reliable detector is challenging. However, both reverse traffic monitoring and RF leakage power tracking schemes are not applicable to all primary network environments. For instance, primary receiver detection through tracking the RF leakage power

A. Khattab et al., *Cognitive Radio Networks: From Theory to Practice*, Analog Circuits and Signal Processing, DOI 10.1007/978-1-4614-4033-8_3,

is only feasible for the detection of TV receivers. Furthermore, both schemes still cannot provide information about the actual interference at the primary receivers. Hence, they cannot help the secondary users determining the appropriate transmission parameters.

Alternatively, cognitive user cooperation could be used to improve the accuracy of spectrum sensing. Collecting the spectrum sensing information from multiple cognitive radio nodes minimizes the uncertainty about the spectral opportunities [12]. However, such cooperative techniques significantly reduce the goodput of the CRN due to the associated overhead traffic required for node cooperation and further complicate the CRN design (e.g., it might necessitates the need for an overlay subnetwork that is only responsible for coordinating the cooperation process).

3.2 Opportunistic Spectrum Access and Sharing

The spectrum access problem is to determine the resources to be used for an upcoming transmission. Such a resource allocation decision includes both the identity of the spectrum to be used alongside a transmission scheme to be used (defined in terms of the transmission power and the modulation rate) and the time instance such a spectrum is available. On the other hand, the spectrum sharing problem considers multiuser scenarios and jointly allocates the available resources among different secondary flows. Due to the close relationship between the two problems they are generally jointly addressed. Several spectrum access and sharing schemes have been proposed for CRNs with the general objective of maximizing the CRN goodput without violating the interference (and consequently, the outage) constraints of the primary networks [2, 3, 13–22]. One way to classify spectrum access and sharing schemes is based on how the resource allocation decisions are made as follows.

3.2.1 Centralized Spectrum Access/Sharing

Such schemes rely on a single centralized entity that collects the spectrum measurements from different secondary nodes and makes the spectrum access decisions and resource allocation decisions for different transmissions. In [15] a spectrum server is utilized to find the optimal schedules that maximize the average sum rate subject to a minimum average rate constraint for the cognitive links using a graph-theoretic approach. The resulting schedules are a collection of transmission modes (i.e., sets of active links) that are time shared in a fashion that is reminiscent of the spatial reuse patterns in cellular networks. In [16], a joint admission control and resource scheduling policy is proposed based on Layapunov optimization techniques. Alternatively, [17] adopts a game-theoretic approach to find the optimal channel assignments and transmission powers/rates. The authors of [17] consider the IEEE 802.22 network model in which a set of base stations are responsible for spectrum access and management and analyze the performance of both cooperative and non-cooperative schemes.

3.2.1.1 Related Standardization Effort: IEEE 802.22 Standard

Despite the recent growth in the wireless communications industry, there still exist large unexploited markets for broadband wireless access in unserved/underserved areas where the wired infrastructure is economically not feasible to deploy. Thus motivated, the IEEE 802.22 standard: "Wireless Regional Area Networks (WRANs): Enabling Rural Broadband Wireless Access Using Cognitive Radio Technology in TV Whitespaces" was initiated to provide reliable and secure broadband wireless connectivity to underserved and unserved communities. The IEEE 802.22 incorporates advanced cognitive radio capabilities including dynamic spectrum access, incumbent database access, accurate geolocation techniques, spectrum sensing, regulatory domain dependent policies, spectrum etiquette, and coexistence to increase the efficiency of spectrum utilization in spectrum currently allocated to, but unused by, the TV broadcast service. In the United States, the abandoned television frequencies are primarily in the upper UHF 700 MHz band, covering TV channels 52–69 (698–806 MHz). In other countries worldwide, the abandoned television channels are in the VHF band. The IEEE 802.22 standard takes advantage of the favorable transmission characteristics of the VHF and UHF TV bands to provide up to 22 Mbps broadband wireless access over a cell radius of up to 100 km (if power is not an issue[1]) without interfering with the reception of existing TV broadcast stations, using the so-called white spaces between the occupied TV channels. While the initial version of the standard does not allow mobility, the physical layer of the IEEE 802.22 standard can support mobility of up to 114 km per hour. The IEEE 802.22 work group was initiated as the first worldwide effort to develop a cognitive radio-based air interface (i.e., the PHY and MAC layer specifications) in November 2004 and the standard was published in July 2011 [23].

The IEEE 802.22 system specifies a fixed point-to-multipoint wireless air interface whereby a Base Station (BS) manages all of the associated Consumer Premise Equipments (CPEs) within the cell as shown in Fig. 3.1. The centralized BS is responsible for the spectrum access decision and the transmission parameters (e.g., modulation, coding, and frequencies of operation) for all of the associated CPEs in both the downlink and uplink directions. The IEEE 802.22 standard adopts a strict master/slave relationship in which the BS acts as the master and the CPEs are the slaves. Such a master/slave approach ensures the protection of incumbent services such as the TV service and wireless microphones. Furthermore, the CPEs

[1]Current specified coverage range is 33 km at 4 W CPE effective isotropically radiated power (EIRP).

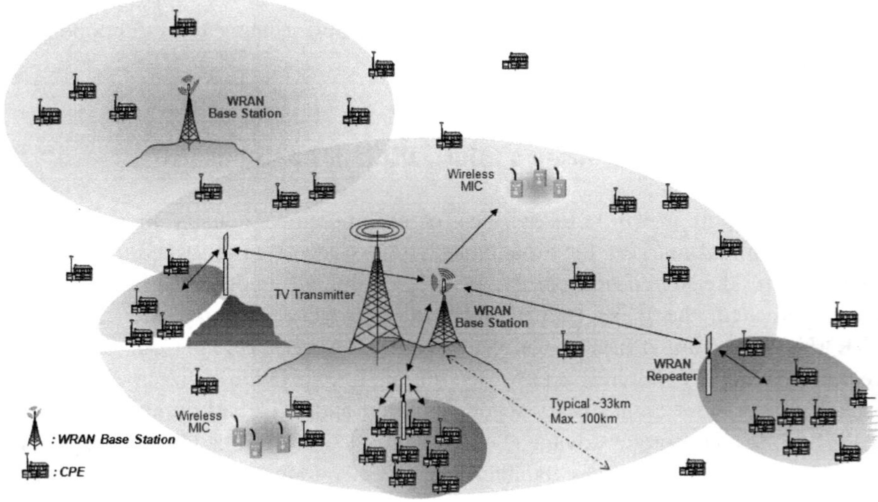

Fig. 3.1 Typical IEEE 802.22 deployment as reported in [24]

are not allowed to transmit before receiving the proper authorization from a BS. In order to improve the accuracy of spectrum sensing and ensure proper incumbent protection, the IEEE 802.22 BS exploit distributed sensing which have various CPEs performing distributed measurement activities. Based on the received information, the BS better decides the best course of action to be taken.

3.2.2 *Distributed Spectrum Access/Sharing*

In a more related context, several distributed spectrum management schemes have been proposed for ad-hoc Cognitive Radio Networks [2, 3, 18–22]. The CRN users individually or jointly decide their channel allocations and transmission powers/rates without a centralized coordinator. For instance, the authors of [18] adopt a decision-theoretic approach and present an analytical framework that integrates the design of spectrum access protocols at the MAC layer, the spectrum sensing at the physical layer and the traffic statistics determined by the application layer of the licensed networks. In [19], different sensing-based opportunistic access schemes were proposed via modeling the primary users as M/G/1 queues and introducing the collision probability and the overlapping time as the metrics for primary user performance. Meanwhile, [20] presents a price-based spectrum management framework for CRNs. The framework models the CRN problem as a non-cooperative game and uses a price-based iterative water-filling algorithm to reach Nash equilibrium. Alternatively, [21] presents a novel cooperative game-theoretic paradigm that allows secondary users to freely optimize the channel utilization for transmitting the primary network data along with their own data. Learning techniques have also been employed to find the optimal resource allocation (time, spectrum, power, and rate) that maximizes the goodput the CRN [2, 3]. In [2], two distributed cooperative learning and allocation schemes were proposed: one that assumes minimal prior knowledge of the secondary user information and the other does not assume such information. The objective of both schemes is to minimize the total regret in distributed learning (or equivalently maximize the CRN goodput). Similarly, [3] utilizes adaptive learning for spectrum probing that is integrated with the resource allocation to maximize the CRN goodput. The authors of [22] propose a scheme based on traditional Carrier Sense Multiple Access with Collision Avoidance (CSMA/CA) MAC that does not rely on the interaction with the licensed networks. Instead, the resource allocation decisions are based on the statistics of the interference over different spectrum bands.

In contrast to all of the aforementioned distributed schemes, the goal of this book is to avoid the use of any inter-flow coordination mechanism. We introduce a probabilistic and non-greedy access mechanism to allow competing secondary flows to efficiently and fairly share the available spectral opportunities without relying on the common control channel for inter-flow communication. Hence, the common control channel is no longer the bottleneck of the CRN nor the single point of failure of the system.

3.2.2.1 Related Standardization Effort: IEEE 1900 Standard

The IEEE Standardization Coordinating Committee 41 (SCC41), "Dynamic Spectrum Access Networks (DySPAN)" targets facilitating the development of the research ideas related to distributed Cognitive Radio Networks into standards to expedite the use of such ideas in commercial products [25]. This standardization effort is based on the IEEE P1900 initiative started in 2004. The DySPAN goals are to:

- Standardize dynamic spectrum access radio systems and networks while focusing on improving the spectrum utilization.
- Develop new dynamic spectrum access techniques and cognitive networking methods as well as the management of radio transmission interference in such a heterogenous environment.
- Coordinate the activities and the deployment of different wireless technologies to share the available spectrum.

Currently, the standard consists of six active work groups that can be summarized as follows.

- **IEEE 1900.1: Definitions and Concepts for DSA: Terminology Relating to Emerging Wireless Networks, System Functionality, and Spectrum Management.** The IEEE 1900.1 standard provides the definitions and explanations of the key concepts in the fields of spectrum management, cognitive radio, policy-defined radio, adaptive radio, software-defined radio, and related technologies. Such a common dictionary of terms is critical to the cognitive radio community to ensure a unified understanding of the used technical terms. The IEEE 1900.1 was approved on September 2008.

- **IEEE 1900.2: Recommended Practice for the Analysis of In-Band and Adjacent Band Interference and Coexistence Between Radio Systems.** Motivated by the emergence of different spectrum management, policy-defined radio, adaptive radio, and software-defined radio systems, the IEEE 1900.2 initiative aims at improving the spectral efficiency and allowing the coexistence of different cognitive radio technologies. The resultant recommended practice will provide the technical guidelines for analyzing the potential for the coexistence—or in contrast the interference—between radio systems operating in the same frequency band or between different frequency bands. The IEEE 1900.2 was approved on July 2008.
- **IEEE 1900.4: Architectural Building Blocks Enabling Network-Device Distributed Decision Making for Optimized Radio Resource Usage in Heterogeneous Wireless Access Networks.** In February 2009, the IEEE has approved the IEEE 1900.4 standard. The purpose of 1900.4 is to address radio resource management, reconfiguration management in composite wireless networks comprising multiple radio access technologies. It addresses the optimization of resources in both fixed and dynamic spectrum access contexts. The IEEE 1900.4 standard defines the architectural building blocks enabling network-device distributed decision making for optimized radio resource usage in heterogeneous wireless networks. The building blocks comprising (1) network resource managers, (2) device resource managers, and (3) the information to be exchanged between the building blocks, for enabling coordinated network-device distributed decision making that will aid in the optimization of radio resource usage, including spectrum access control, in heterogeneous wireless access networks are defined. The IEEE 1900.4 defines a policy-based management framework for decision making. The IEEE 1900.4 dynamic policy-based approach makes it well suited to the changing and non-homogeneous white spaces context. The IEEE then started two work groups based on the 1900.4 standard: the IEEE 1900.4a and the IEEE 1900.4.1. The IEEE 1900.4a amendment of the original IEEE 1900.4 standard is concerned with architecture and interfaces for dynamic spectrum access networks in white space frequency bands (i.e., unused wireless spectrum bands). The final standard would include the definition of the new devices and interfaces required for operating over the white space frequency bands in addition to IEEE 1900.4 entities and interfaces. While the white space spectrum might vary depending on the country or the region, the IEEE 1900.4a standard provides a common management system that is independent of the radio interface of choice. Such a unique management system enables the white spaces devices to comply with the regulation irrespective of the geographic location and time. Hence, the protection of broadcast systems will be absolutely reliable. On the other hand, the 1900.4.1, "Standard for Interfaces and Protocols Enabling Distributed Decision Making

for Optimized Radio Resource Usage in Heterogeneous Wireless Networks" is responsible for providing the details of the interfaces and service access points defined in the IEEE 1900.4 standard in order to ensure interoperability between network side and terminal side components of the IEEE 1900.4 system.

- **IEEE 1900.5: Policy Language and Policy Architectures for Managing Cognitive Radio for Dynamic Spectrum Access Applications.** This standard defines the policy language and the associated architecture requirements for interoperable, vendor-independent control of Dynamic Spectrum Access functionality and behavior in radio systems and wireless networks. In other words, it defines the relationship of policy language and architecture to guarantee consistency between the regulator, the operator, the user, and the network equipment manufacturer. The work on the IEEE 1900.5 standard started in August 2008 and the draft standard was published in the end of 2011.
- **IEEE 1900.6: Spectrum Sensing Interfaces and Data Structures for Dynamic Spectrum Access and other Advanced Radio Communication Systems.** The objective of the IEEE 1900.6 work group is to define the interfaces and the data structures required to exchange sensing-related information in order to increase the interoperability between sensors and their clients developed by different manufacturers. The logical interface and supporting data structures are defined abstractly without constraining the sensing technology, client design, or data link between sensor and client. It also defines the involved entities and the exchanged parameters in this process. It further elaborates on the service access points, service primitives, as well as generic procedures used to realize this information exchange, are defined by this standard. The IEEE 1900.6 standard was published in April 2011. In June 2011, the IEEE 1900.6 work group started working on the IEEE 1900.6a amendment: Procedures, Protocols and Data Archive Enhanced Interfaces. The goal of this amendment is to provide the specifications that allow the integration of 1900.6 based distributed sensing systems into existing and future dynamic spectrum access networks. It enables existing legacy communication systems to benefit from adopting the IEEE 1900.6 interface as an add-on to these systems and to claim standard conformance for an implementation of the interface. Furthermore, this amendment facilitates the sharing of spectrum sensing data and other relevant data among 1900.6 based entities and external data archives.
- **IEEE 1900.7: Radio Interface for White Space Dynamic Spectrum Access Radio Systems Supporting Fixed and Mobile Operation.** This work group targets specifying a radio interface including the MAC sublayer(s) and PHY layer(s) of white space dynamic spectrum access radio systems supporting fixed and mobile operation in white space frequency bands, while avoiding causing harmful interference to incumbent users in these frequency bands. The goal is to provide means to support P1900.4a for white space management and P1900.6 to obtain and exchange sensing related information (spectrum sensing and geolocation information). The standard is not approved yet.

It is worth mentioning that the IEEE P1900.3 work group, "Dependability and Evaluation of Regulatory Compliance for Radio Systems with Dynamic Spectrum

Access", was dismantled in 2008 due to the lack of volunteers. The original goal of that standardization effort was to provide a set of tests and evaluation methods to be used in regulatory compliance testing of cognitive radio terminals.

3.3 Opportunistic Spectrum Access Implementations

The implementation issues and challenges of Opportunistic Spectrum Access have received far less research interest compared to the significant theoretical interest in such a promising communication paradigm. Here, we briefly overview the existing OSA implementation efforts.

3.3.1 SDR-Based OSA Implementations

In his seminal work [26, 27], Mitola defined the Software-Defined Radio (SDR) as the ideal environment for implementing cognitive radios due to the seamless flexibility and controllability of such platforms. A SDR platform relies on a low-cost Universal Software Radio Peripheral (USRP) [28] interfaced to a general purpose computer that runs the software development environment: the GNU radio [29]. Thus motivated, several OSA implementation frameworks have been developed to provide the software libraries and environments that enable the fast composition of OSA protocols. Examples include the Cognitive Radio Open Source System (CROSS) [30], the Papyrus software platform [31], and the Sora platform [32]. However, such SDR-based platforms have a major shortcoming in their response time due to the communication latency between the USRP and the general purpose processor (with the exception of Sora [32] which uses external hardware including dedicated Central Processing Unit (CPU) cores and large low-latency caches to address such a limitation). A comprehensive discussion of the shortcomings of SDR-based platforms is presented in [33].

3.3.2 Commodity Hardware-Based OSA Implementations

Several attempts have been made to use commodity hardware to implement cognitive radio nodes. For example [34, 35], and [36] present cognitive radio platforms that use commodity IEEE 802.11 hardware. While such an implementation approach provides a low-cost solution, it has a limited room for reconfigurability and customization since such an IEEE 802.11 hardware is restricted in providing accessability to the underlying physical layer parameters. This contradicts with one of the two main requirements of a cognitive radio defined by Mitola in [26, 27]. Furthermore, such platforms do not allow for clean-slate OSA designs since only derivatives of the IEEE 802.11 medium access mechanism can be implemented. Alternatively, [37] adopts a different approach in which a suite of predefined stand-alone medium access control protocols are implemented using of an IEEE 802.11 hardware. Furthermore, [37] allows for the control of some physical layer parameters. However, such an approach has limited capabilities as it can only choose the approximate "best-fit" MAC approach from a limited number of the implemented protocols. Thus motivated, the author of [38] proposed a modular approach to break down the different medium access approaches into a few modular designs. However, both the platforms proposed in [37] and [38] lack the flexibility of SDR-based platforms and are incapable of satisfying the requirements of fully-capable CRNs.

3.3.3 FPGA-Based OSA Implementations

In contrast, Field Programmable Gate Array (FPGA)-based platforms provide both the flexibility of SDR-based platforms and the practical performance of hardware-based platforms. However, such remarkable performance of FPGA-based platforms comes at the expense of increased complexity and cost. An example FPGA-based prototype is the network-centric cognitive radio (WiNC2R) board [39]. The WiNC2R platform is equipped with a tri-band radio front-end suitable for operation at the 700 MHz, 2.4 GHz and 5 GHz bands. Likewise, DARPA is currently in the process of developing a hand-held FPGA-based cognitive radio terminal covering the frequency range between 900 MHz and 6 GHz as a part of the Wireless Networks after Next (WNaN) program [40]. One of the fundamental objectives of the WNaN program is to reduce the cost of the transceivers to the point where a sophisticated, multi-transceiver cognitive radio can achieve a cost point below that of conventional technology [41]. Unlike the above two FPGA platforms specifically designed for cognitive radio applications, general purpose wireless FPGA platforms can be used for the implementation of Opportunistic Spectrum Access protocols. For example, the authors of [42] present a framework of implementing OSA protocols using the general purpose Wireless open-Access Research Platform (WARP) [43]. However, that framework is a derivative of the IEEE 802.11 medium access approach. In contrast, this book presents the first implementation and performance evaluation of clean-slate Opportunistic Spectrum Access approaches using WARP.

References

1. Yucek, T., Arslan, H.: A survey of spectrum sensing algorithms for cognitive radio applications. IEEE Comm. Surv. Tutorials **11**(1), 116–130 (2009)
2. Anandkumar, A., Michael, N., Tang, A.: Opportunistic spectrum access with multiple users: Learning under competition. In: Proceedings of IEEE INFOCOM 2010, San Deigo, CA (2010)

3. Chaporkar, P., Proutiere, A., Asnani, H.: Learning to optimally exploit multi-channel diversity in wireless systems. In: Proceedings of IEEE INFOCOM 2010, San Diego, CA (2010)
4. Bhandari, V., Vaidya, N.H.: Capacity of multi-channel wireless networks with random (c, f) assignment. In: Proceedings of ACM Mobihoc 2007, Montreal, Canada (2007)
5. Tian, Z., Giannakis, G.: Compressed sensing for wideband cognitive radios. In: Proceedings of IEEE ICASSP, Honolulu, HI (2007)
6. Liu, H., Krishnamachari, B.: Randomized strategies for multi-user multi-channel opportunity sensing. In: Proceedings of IEEE CCNC Cognitive Radio Networks Workshop, Las Vegas, NV (2008)
7. Liang, Z., Liu, W., Zhou, P., Gao, F.: Randomized multi-user strategy for spectrum sharing in opportunistic spectrum access network. In: Proceedings of IEEE ICC Workshops, Beijing, China (2008)
8. Ahmad, B.I., Tarczynski, A.: Reliable wideband multichannel spectrum sensing using randomized sampling schemes. Signal Process. **90**(7), 2232–2242 (2010)
9. Lapiccirella, F.E., Ding, Z., Liu, X.: Cognitive spectrum access control based on intrinsic primary ARQ information. In: Proceedings of IEEE ICC 2010, Cape Town, South Africa (2010)
10. Lee, C.H. Haenggi, M.: Delay analysis of spatio-temporal channel access for cognitive networks. In: Proceedings of IEEE ICC 2011, Kyoto, Japan (2011)
11. Wild, B., Ramchandran, K.: Detecting primary receivers for cognitive radio applications. In: Proceedings of IEEE DySPAN 2005, Baltimore, MD (2005)
12. Mishra, S.M., Sahai, A., Brodersen, R.W.: Cooperative sensing among cognitive radios. In: Proceedings of IEEE ICC 2006, Istanbul, Turkey (2006)
13. Salameh, H.B., Krunz, M.: Channel access protocols for multihop opportunistic networks: challenges and recent developments. IEEE Networks **23**(4), 14–19 (2009)
14. Akyildiz, I.F., Lee, W.Y., Chowdhury, K.R.: CRAHNs: Cognitive radio ad hoc networks. Ad Hoc Networks (Elsevier) **7**(5), 810–836 (2009)
15. Raman, C., Yates, R.D., Mandayam, N.B.: Scheduling variable rate links via a spectrum server. In: Proceedings of IEEE DySPAN 2005, Baltimore, MD (2005)
16. Lotfinezhad, M., Liang, B., Sousa, E.S.: Optimal control of constrained cognitive radio networks with dynamic population size. In: Proceedings of IEEE INFOCOM 2010, San Diego, CA (2010)
17. Hosseinabadi, G., Manshaei, M.H., Hubaux, J.P.: Spectrum sharing games of infrastructure-based cognitive radio networks. Tech. rep. http://infoscience.epfl.ch/record/128112?ln=en (2008)
18. Zhao, Q., Tong, L., Swami, A., Chen, Y.: Decentralized cognitive MAC for opportunistic spectrum access in ad hoc networks: A POMPD framework. IEEE J. Sel. Area. Comm. **25**(3), 589–600 (2007)
19. Huang, S., Liu, X., Ding, Z.: Opportunistic spectrum access in cognitive radio networks. In: Proceedings of IEEE INFOCOM 2008, Phoenix, AZ (2008)
20. Wang, F., Krunz, M., Cui, S.: Price-based spectrum management in cognitive radio networks. IEEE J. Sel. Top. Signal Process. **2**(1), 74–87 (2008)
21. Xu, H., Li, B.: Efficient resource allocation with flexible channel cooperation in OFDMA cognitive radio networks. In: Proceedings of IEEE INFOCOM 2010, San Diego, CA (2010)
22. Salameh, H.B., Krunz, M., Younis, O.: MAC protocol for opportunistic cognitive radio networks with soft guarantees. IEEE Trans. Mobile Comput. **8**(10), 1339–1352 (2009)
23. IEEE Working Group on Wireless Regional Area Networks: Enabling rural broadband wireless access using cognitive radio technology in TV whitespaces. http://www.ieee802.org/22/. Accessed 25 July 2012
24. Benko, J., Chang, S.Y., Cheong, Y.C., Cordeiro, C., Gao, W., Hu, W., Khalona, R., Kim, C.J., Kim, H.S., Kuffner, S., Laskar, J., Liang, Y.C., Sofer, E.: IEEE802.22-06/0069r2: Draft PHY/MAC specification for IEEE 802.22 (2006)
25. IEEE DySPAN Standards Committee: Dynamic Spectrum Access Networks (DySPAN). http://www.dyspan-sc.org/. Accessed 25 July 2012

26. Mitola III, J.: Cognitive radio: An integrated agent architecture for software defined radio. Ph.D. thesis, KTH Royal Institute of Technology (2000)
27. Mitola III, J.: Cognitive radio for flexible mobile multimedia communication. In: Proceedings of IEEE International Workshop on Mbile Multimedia Communications (MoMuC), San Diego, CA (1999)
28. Ettus Research LLC: http://www.ettus.com/. Accessed 25 July 2012
29. GNU Radio: http://gnuradio.org/redmine/projects/gnuradio/wiki. Accessed 25 July 2012
30. CROSS: Cognitive Radio Open Source System. http://cornet.wireless.vt.edu/trac/wiki/Cross. Accessed 25 July 2012
31. Yang, L., Zhang, Z., Hou, W., Zhao, B.Y., Zheng, H.: Papyrus: A software platform for distributed dynamic spectrum sharing using SDRs. ACM SIGCOMM Comput. Comm. Rev. **41**, 31–37 (2011)
32. Tan, K., Zhang, J., Fang, J., Liu, H., Ye, Y., Wang, S., Zhang, Y., Wu, H., Wang, W., Voelker, G.M.: Sora: High-performance software radio using general-purpose multi-core processors. Comm. ACM **54**, 99–107 (2011)
33. Nychis, G., Hottelier, T., Yang, Z., Seshan, S., Steenkiste, P.: Enabling MAC protocol implementations on software-defined radios. In: Proceedings of USENIX symposium on NSDI, Boston, MA (2009)
34. Sharma, A., Belding, E.M.: FreeMAC: Framework for multi-channel MAC development on 802.11 hardware. In: Proceedings of ACM PRESTO'08 Workshop, Seattle, WA (2008)
35. Lu, M.H., Steenkiste, P., Chen, T.: FlexMAC: A wireless protocol development and evaluation platform based on commodity hardware. In: Proceedings of ACM WiNTECH 2008, San Francisco, CA (2008)
36. Sharma, A., Tiwari, M., Zheng, H.: MadMAC: Building a reconfigurable radio testbed using commodity 802.11 hardware. In: Proceedings of IEEE SECON WSDR, Reston, VA (2006)
37. Doerr, C., Neufeld, M., Fifield, J., Weingart, T., Sicker, D., Grunwald, D.: MultiMAC - an adaptive MAC framework for dynamic radio networking. In: Proceedings of IEEE DySPAN 2005, Baltimore, MD (2005)
38. Messerschmitt, D.G.: Rethinking components: From hardware and software to systems. Proc. IEEE **95**, 1473–1496 (2007)
39. Miljanic, Z., Seskar, I., Le, K., Raychaudhuri, D.: The WINLAB network centric cognitive radio platform - WiNC2R. In: Proceedings of CrownComm 2007, Orlando, FL (2007)
40. Marshall, P.: Extending the reach of cognitive radio. Proc. IEEE **97**, 612–625 (2009)
41. DARPA's Wieless Network after Next Project: http://www.darpa.mil/Our_Work/STO/Programs/Wireless_Network_after_Next_(WNAN).aspx. Accessed 25 July 2012
42. Ansari, J., Zhang, X., Achtzehn, A., Petrova, M., Mahonen, P.: A flexible mac development framework for cognitive radio systems. In: Wireless Communications and Networking Conference (WCNC), 2011 IEEE, Quintana Roo, Mexico (2011)
43. Rice University WARP Project: http://warp.rice.edu. Accessed 25 July 2012

Chapter 4
Opportunistic Spectrum Access Challenges in Distributed Cognitive Radio Networks

4.1 Network Model

4.1.1 Primary Network Model

We consider a wireless spectrum consisting of N non-overlapping frequency bands each with bandwidth B_i. We assume N distinct Primary Radio Networks (PRNs) licensed to operate in these N bands.[1] All of the N PRNs are geographically collocated. The maximum transmission power of the ith PRN is $P_{PU}^{(i)}$. The PRN user distributions are modeled as *i.i.d.* homogeneous Poisson random processes with parameters ρ_i representing the user density of the ith PRN. A Primary User (PU) in the ith PRN is modeled as an ON/OFF source with activity factor α_i defined as the fraction of time the user is ON. PRNs are non-intrusive and operate as they are the sole users of their licensed spectrum. PUs do not provide any type of cooperation with the underlying secondary Cognitive Radio Network. However, each PRN defines the maximum permissible interference margin from the secondary Cognitive Radio Network. We denote such a power mask of the ith PRN (and consequently the ith band) as $P_{mask}^{(i)}$. We adopt a statistical model that ensures that the cumulative interference from the secondary user activities does not exceed $P_{mask}^{(i)}$ with probability β, thereby, providing a stochastic guarantee on the performance of PUs.

[1]A PRN can be licensed to use multiple contiguous or non-contiguous bands. However, our generalized assumption of a different PRN per band can be easily extended to capture such situations by dividing such a multi-band PRN into multiple virtual PRNs.

A. Khattab et al., *Cognitive Radio Networks: From Theory to Practice*,
Analog Circuits and Signal Processing, DOI 10.1007/978-1-4614-4033-8_4,

4.1.2 Secondary Network Model

We consider a single distributed secondary Cognitive Radio Network (CRN) that is geographically collocated with the N PRNs. Transmissions within different PRNs and the CRN can start at any arbitrary time instant (i.e., we do not assume a time-slotted system). The unlicensed users of the CRN can opportunistically access any of the N non-overlapping bands one band at a given time. A Secondary User (SU) is equipped with a single cognitive radio transceiver that can be tuned to transmit over any of the N bands. We assume the transceiver has a narrow-band sensing capability. That is a SU transceiver can only sense a single frequency band at a time. While not optimal compared to wide-band sensing, narrow-band spectrum sensing relaxes the hardware complexity and the power consumption of SU terminals (especially for low-cost battery-powered devices). SUs are of lower priority with respect to spectrum access compared to the band's licensed PUs. The secondary user density is ρ_{SU}. We consider a multiuser CRN environment in which one or more SUs can transmit over a given band once a spectral opportunity is inferred (i.e., the sensed cumulative interference power on the ith band is less than $P_{mask}^{(i)}$). We denote the transmission power of the jth SU over the ith band as $P_{SU_j}^{(i)}$ and the corresponding transmission rate as $r_{SU_j}^{(i)}$. Both $P_{SU_j}^{(i)}$ and $r_{SU_j}^{(i)}$ are fixed throughout a packet transmission. A SU can choose its rate from a finite set of available rates $R_1 < R_2 < \ldots < R_m$. Each rate R_i has a corresponding distinct transmission power $P_1 < P_2 < \ldots < P_m$. The powers P_i's are such that the transmission range is fixed irrespective of the used rate. Thus, the following relationship holds for any pair of rates

$$\frac{P_i}{P_j} = \frac{2^{R_i} - 1}{2^{R_j} - 1}, \qquad \forall i \neq j \tag{4.1}$$

due to the logarithmic relationship between the rate and power regardless of the used physical layer scheme [1]. A secondary sender-receiver pair coordinates its spectrum selection and transmission policy using a dedicated common control channel in the unlicensed band. Unlike prior work, the common control channel is not used for any sort of inter-flow coordination.

4.2 Cognitive Radio Network Challenges

Here, we review the unsolved challenges of CRNs to be addressed in the remainder of this book. Namely, the inability of current spectrum sensing schemes to accurately assess spectral opportunities based on local interference measurements and the multi-flow coordination problem in distributed CRNs.

4.2.1 *Spectrum Sensing Challenges*

There exists many factors that limit the performance of spectrum sensing and necessitate spectrum measurements to be inaccurate.

4.2.1.1 Cognitive Radio Hardware Limitations

The first challenge in spectrum sensing comes from the cognitive transceiver hardware that is required to achieve sufficiently high sensitivity for a wide spectrum range (e.g., multiple gigahertz) while accurately detecting diverse and frequency-dependent primary signals at different received power levels. This places severe requirements on the sensitivity, linearity, and dynamic range of the circuitry in the RF front-end, and more specifically, the antennas, power amplifiers and the analog-to-digital conversion units [2–4]. Furthermore, the processing power requirements of the signal processing units that analyze the sensed spectrum are very high in order for the cognitive radio to make a decision with a relatively low delay. Note that the environment is dynamic and the interference is modulated both by the bursty traffic of primary users and channel fading.

Furthermore, the finite sensing duration also limits the accuracy of spectrum sensing. Given a limited sensing period, only a certain sensing accuracy can be guaranteed. As the sensing duration increases, the reliability of the sensing information might increase. However, longer spectrum sensing windows are not necessarily useful since the environment is dynamic and the energy on a given channel is modulated both by the bursty traffic, asynchronous initiation and termination of packet transmissions, and channel fading [2].

4.2.1.2 Listen-Before-Talk Spectrum Sensing Dilemma

The most important factor that significantly tackles down the performance of spectrum sensing is that most of the existing techniques adopt some form of the traditional listen-before-talk strategy to detect the activities of the primary transmitters. However, listen-before-talk strategies do not provide information about the radio activities at the primary receivers. Recall that CRNs are required to not cause interference at the primary receivers to satisfy the primary network outage constraints. Currently, there does not exist any practical way that allows cognitive nodes to measure the interference at nearby primary network receivers [2, 3, 5] since the primary users are passive and do not interact or share information with the secondary users. Therefore, interference measurements based on local observations at SUs are inaccurate.

Such erroneous spectrum measurements cause the SUs to mistakenly infer spectral opportunities or miss spectral opportunities as is the case in the scenarios depicted in Fig. 4.1a, b, respectively. In the scenario depicted in Fig. 4.1a, the

a

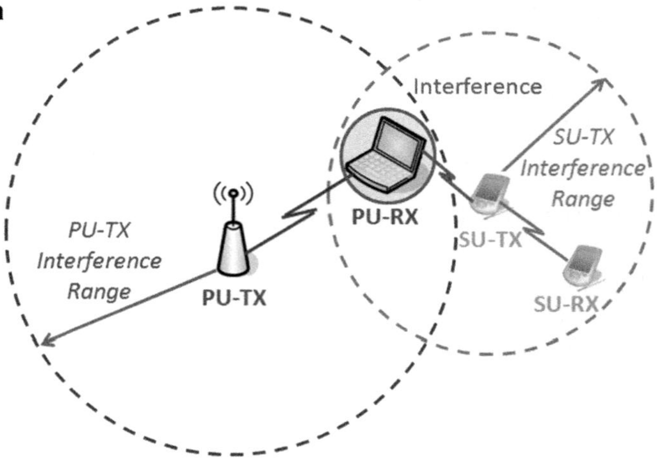

Hidden primary sender scenario as an example of spectral opportunity miss-prediction. The cognitive radio sender will mistakenly exploit such a spectrum due to its inability to infer the existence of the far away primary transmitter. Hence, the cognitive radio transmission will interfere with the primary transmission at the nearby primary receiver.

b

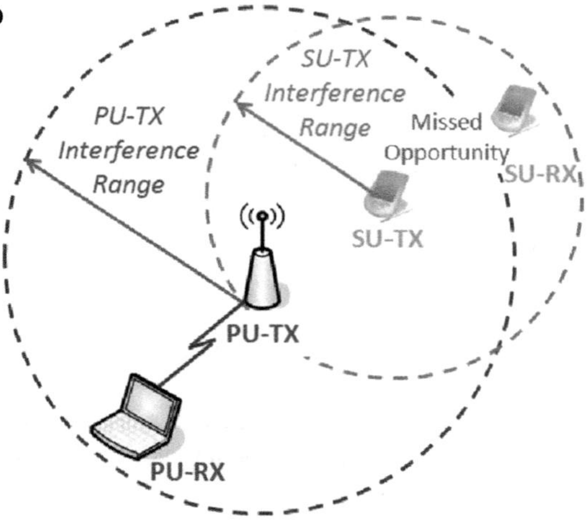

Exposed primary sender as an example of missing a spectral opportunity. The cognitive radio sender will mistakenly not exploit such a spectrum due to the existence of a nearby primary transmitter. However, using such a spectrum would have not interfered with the primary transmission at the far away primary receiver.

Fig. 4.1 The lack of a practical way to infer the interference at primary receivers causes spectrum miss-predictions (**a**) and missing spectral opportunities (**b**)

received power at the secondary transmitter (SU-TX) from the primary transmitter (PU-TX) is below the SU receive sensitivity, and hence, the SU-TX infers the existence of a spectral opportunity at the spectrum used by the PRN due to its interference measurement of that spectrum band. Therefore, the SU-TX will start its transmission. However, such a SU transmission will interfere with the reception of the PRN data since the SU-TX decision did not take into account its impact on the undetected nearby primary receiver (PU-RX). We refer to such a scenario as opportunity miss-prediction. In the scenario shown in Fig. 4.1b, the SU-TX decides not to use the spectrum used by the PU-TX based on its measurement of a strong power received from the nearby PU-TX. However, the SU-TX transmission would not have disturbed the PRN transmission if the SU was to transmit. We refer to such a situation as a missed opportunity scenario. Note the analogy of the miss-prediction and missed opportunity scenarios with the hidden terminal and exposed terminal problems in traditional ad-hoc networks. However, the use of the request-to-send/clear-to-send (RTS/CTS) handshake does not alleviate the miss-prediction and missed-opportunity problems as the case with the hidden and exposed terminal problems.

Recent works exploited the bidirectional nature of some primary networks to enable SUs to infer the existence or the absence of a neighboring primary receiver [6, 7]. However, such schemes do not provide a way to measure the cumulative interference at the primary receiver. Another potential approach to collect information about nearby primary receivers is to detect the local oscillator leakage power emitted by the RF front-end of the primary receivers. Since such leakage power is typically weak, implementing a reliable detector is challenging. As per existing transceiver technologies, primary receiver detection through tracking the RF leakage power is only feasible for the detection of TV receivers. A more detailed discussion of the related work was presented in Chap. 3. Alternatively, cognitive user cooperation could be used to improve the accuracy of spectrum sensing. Collecting the spectrum sensing information from multiple cognitive radio nodes minimizes the uncertainty about the spectral opportunities. However, such cooperative techniques significantly reduce the goodput of the CRN due to the associated overhead traffic required for node cooperation and further complicate the CRN design (e.g., it might necessitates the need for an overlay subnetwork that is only responsible for coordinating the cooperation process). We conclude that there does not exist a practical—yet accurate—way to measure the interfere at the primary receivers based on today's sensing technologies.

4.2.2 *Distributed Spectrum Sharing Problem*

In distributed multiuser CRN, the coordination between multiple secondary users is a major challenge. If legacy MAC protocols designed for traditional ad-hoc networks were to be used in CRNs, all secondary users that infer a spectral opportunity will *greedily* attempt to exploit the sensed opportunity. Recall that legacy MACs often adopt greedy strategies that try to best utilize a channel access (e.g., by using the highest transmission rate or choosing the best channel). Such greedy approaches deteriorate the goodput performance of a CRN as the number of SUs increases due to increased blocking probability [3,5]. Furthermore, such greedy MACs are known to suffer from unfairness problems that can cause some secondary sender-receiver pairs to dominate other pairs. Several distributed cooperative MAC approaches have been recently developed for CRNs [8–10]. However, such distributed schemes rely on the explicit coordination between different flows which is a main challenge in CRNs as it requires gathering and distributing spectrum information across the CRN and/or synchronizing the activities of different flows. The use of a common control channel for inter-flow coordination (as well as for the coordination between a sender and its respective receiver) makes it the bottleneck of a CRN and the single point of failure for the entire system [3,5].

4.3 Book Goals

In the remainder of this book, our objectives are to develop, optimize, and realize a practical spectrum management scheme for distributed Cognitive Radio Networks that

1. Counters the *unavoidable* inaccuracies in spectrum measurements and their consequent negative impact on the CRN and PRNs performance.
2. Allows secondary users to fairly share spectral opportunities without explicit inter-flow coordination.

Given the limited capabilities of existing radio technologies, the proposed scheme should relax the hardware requirements of the cognitive transceivers. Our main idea is to use randomized sensing combined with a probabilistic and non-greedy transmission policy to counter the inaccuracies in spectrum sensing and prevent an individual secondary flow from exclusively exploiting an available opportunity. We address the following two open questions. First, given that a secondary sender does not apriori know the impact of its transmission on nearby primary receivers, how aggressive/conservative a secondary sender should/should not be to alleviate spectral miss predictions and missed opportunities. Second, how non-greedy spectrum access can allow multiple secondary users to share spectral opportunities without explicit information sharing.

Our goals can be summarized as follows. First, we introduce our probabilistic Opportunistic Spectrum Access framework and discuss the details of its protocol implementation. Second, we analytically model the proposed OSA scheme in order to optimize its performance by computing the optimal values of its parameters. Third, we use large-scale simulations to evaluate the performance of the proposed scheme compared to the wide set of existing OSA that are based on theoretical assumptions—without regard to the practical constraints—in arbitrary ad-hoc networks. Finally, we implement our scheme on the state-of-the-art of wireless hardware platforms to assess its performance in real-life scenarios.

References

1. Rappaport, T.: Wireless communications, principles & practice. Prentice Hall, Englewood Cliffs (1996)
2. Yucek, T., Arslan, H.: A survey of spectrum sensing algorithms for cognitive radio applications. IEEE Comm. Surv. Tutorials **11**(1), 116–130 (2009)
3. Akyildiz, I.F., Lee, W.Y., Chowdhury, K.R.: CRAHNs: Cognitive radio ad hoc networks. Ad Hoc Networks (Elsevier) **7**(5), 810–836 (2009)
4. Cabric, D., Brodersen, R.W.: Physical layer design issues unique to cognitive radio systems. In: Proceedings of IEEE PIMRC, Berlin, Germany (2005)
5. Salameh, H.B., Krunz, M.: Channel access protocols for multihop opportunistic networks: Challenges and recent developments. IEEE Networks **23**(4), 14–19 (2009)
6. Lapiccirella, F.E., Ding, Z., Liu, X.: Cognitive spectrum access control based on intrinsic primary ARQ information. In: Proceedings of IEEE ICC 2010, Cape Town, South Africa (2010)
7. Lee, C.H., Haenggi, M.: Delay analysis of spatio-temporal channel access for cognitive networks. In: Proceedings of IEEE ICC 2011, Kyoto, Japan (2011)
8. Zhao, Q., Tong, L., Swami, A., Chen, Y.: Decentralized cognitive MAC for opportunistic spectrum access in ad hoc networks: A POMPD framework. IEEE J. Sel. Area. Comm. **25**(3), 589–600 (2007)
9. Wang, F., Krunz, M., Cui, S.: Price-based spectrum management in cognitive radio networks. IEEE J. Sel. Top. Signal Process. **2**(1), 74–87 (2008)
10. Salameh, H.B., Krunz, M., Younis, O.: MAC protocol for opportunistic cognitive radio networks with soft guarantees. IEEE Trans. Mobile Comput. **8**(10), 1339–1352 (2009)

Chapter 5
Rate-Adaptive Probabilistic Approach for Opportunistic Spectrum Access

5.1 RAP Framework

The proposed RAP framework has two main components: The randomized spectrum selection component that relaxes the spectrum sensing hardware requirements, combined with the rate-adaptive probabilistic transmission policy which probabilistically: (1) allows secondary senders to better explore the spectral opportunities regardless of the inaccuracy of spectrum sensing, and (2) enables multiple secondary flows to share the available opportunities in a distributed manner without explicit inter-flow coordination.

5.1.1 Coordinated Random Spectrum Selection

As we explained in Chap. 4, secondary senders are unable to apriori assess the impact of their transmissions on nearby primary receivers based on the PU interference in the vicinity of the cognitive radio sender. The resultant inaccuracies cause a secondary transmitter to miss-infer the existence or the lack of spectral opportunities, which lead to wrong spectral access decisions. We propose a probabilistic scheme for spectrum sensing in order to relax the constraints on the spectrum sensing hardware and counter the inherent inaccuracy in assessing the spectral opportunities only based on local SU measurements without PU cooperation. Our spectrum sensing scheme is based on the following two ideas.

5.1.1.1 Randomized Spectrum Selection

A secondary transmitter (SU-TX) randomly selects a spectrum to probe for an upcoming transmission (if there does not exist a preferred spectrum that recently carried out a successful transmission). Due to the inability of a secondary sender

A. Khattab et al., *Cognitive Radio Networks: From Theory to Practice*,
Analog Circuits and Signal Processing, DOI 10.1007/978-1-4614-4033-8_5,

to accurately assess the impact of its transmission on ongoing PU transmissions, a secondary sender can choose any spectrum with *equal probability* for an upcoming transmission. Prior work used randomized spectrum sensing to spread multiple SUs over different spectrum bands [1, 2]. However, such schemes require the exact apriori knowledge of the statistics of the activities of primary users and the number of competing SUs in order to compute the probability of sensing a particular spectrum. In contrast, we use randomization to address the inherent inaccuracy of interference measurements given the limitation of SUs being unable to infer the interference at primary receivers. Furthermore, using such randomized spectrum sensing alleviates the need for wide-band sensing, and hence, relaxes the cognitive radio requirements.

5.1.1.2 Coordinated Sender-Receiver Sensing

The interference at the sender and receiver of a SU flow are typically different in distributed ad-hoc environments (in which nodes are exposed to different parts of the network). Therefore, a spectrum access decision must be based on the view of the spectrum at both endpoints of a given transmission (not only on the sender's view of the spectrum as the case with traditional listen-before-talk MAC protocols). Hence, the RAP framework has the secondary receiver (SU-RX) also measure the interference over the secondary-sender-selected spectrum. Given the interference measurements of the selected spectrum at both the SU-TX and SU-RX, four scenarios arise. In the first scenario, both measurements indicate low interference (i.e., the cumulative interference is below the power mask). We refer to such a scenario as a *clear spectral opportunity*. The second scenario is when the SU-TX is experiencing strong interference (i.e., the cumulative interference exceeds the power mask) and the SU-RX is experiencing low interference. We refer to such scenario as an *unclear spectral opportunity*. The other two scenarios are when the spectrum measurement at the SU-RX indicates high interference levels. In such scenarios, the SU-RX will not be able to correctly receive data over the randomly selected spectrum. The RAP framework avoids unnecessary usage of such a spectrum band by having the SU-TX randomly selecting a new spectrum.

Figure 5.1 illustrates the coordinated random spectrum selection in some scenarios. In the first secondary transmission attempt, the secondary sender randomly picks spectrum i. After measuring spectrum i, it sends a Spectrum Request (SR) packet over the common control channel to its receiver. The receiver does not respond with a Spectrum Grant (SG) packet since spectrum i is currently used by the PRN in its vicinity based on its local spectrum measurement. After a timeout period, the secondary sender randomly picks a new spectrum j and sends a SR packet over the common control channel. The secondary receiver measures the PRN activity in its vicinity and concludes that this spectrum is available. Hence, it responds back with an SG packet. Upon receiving the SG packet, the SU-TX sends a data packet on spectrum j. Then, the SU-RX acknowledges the correct reception of that packet.

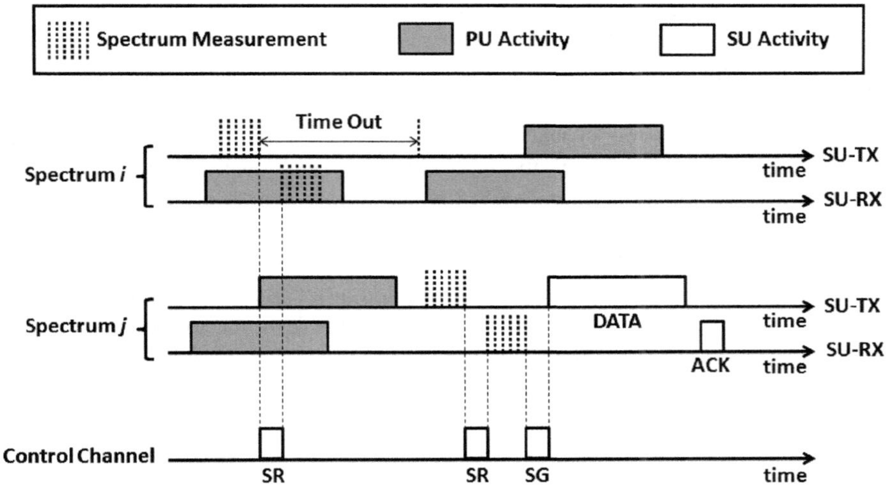

Fig. 5.1 Timeline of the RAP framework operation in some spectrum scenarios illustrating the secondary sender-receiver coordination mechanism

5.1.2 *Rate-Adaptive Probabilistic Transmission*

Even with the spectrum measurements at both the SU-TX and SU-RX, the decision of whether or not to use the sensed spectrum cannot be accurate. We propose the following probabilistic spectrum access scheme which is:

- Conservative and non-greedy in exploiting clear spectral opportunities, and hence, it probabilistically reduces the PRN outages due to spectral miss-predictions while allowing multiple secondary flows to exploit a given spectral opportunity.
- Probabilistically non-conservative in exploiting unclear spectral opportunities in order to reduce the CRN goodput degradation due to spectral missed opportunities.

5.1.2.1 Clear Spectral Opportunity

In clear spectral opportunity scenarios, the RAP framework exploits the sender-selected spectrum at the maximum permissible power/rate only with a certain probability p (since a SU-TX does not know for sure if its transmission will interfere with any ongoing primary receptions or not). Besides, such a non-greedy medium access approach does not allow a SU-TX to fully utilize the available capacity of a given spectral opportunity (i.e., a SU-TX does not transmit with the highest possible power and rate). Instead, a SU-TX probabilistically leaves a capacity margin by using a lower power/rate with probability $(1 - p)$. Hence, if there exists a neighboring SU transmission it can exploit such a capacity margin to announce its presence. Consequently, different SU transmissions adjust their powers and rates to share such an opportunity.

While potentially degrading the CRN goodput, the use of a low power/rate transmission reduces the probability of intercepting ongoing unidentified PRN transmissions since the lower the rate the lower its power. Starting from the minimum values, a SU-TX increases the rate and power used with probability $(1 - p)$ to the next higher values upon a successful transmission until either the second highest values are reached or a transmission failure occurs. The purpose of the former condition is to not sacrifice the goodput of the CRN if there does not exist any nearby SU transmissions by gradually shrinking the unutilized capacity margin. Meanwhile, if a nearby secondary transmission decides to explore the same spectrum it will cause the high rate transmission to fail. In this case, our scheme will have a SU-TX reverting to the lowest power/rate for future transmissions. Low power/rate communication scheme are more robust to interference that cannot be explicitly nulled out [3]. It was shown that multiple low power and low rate transmission successfully coexist without explicit interference suppression [4].

5.1.2.2 Unclear Spectral Opportunity

In unclear spectral opportunity scenarios, the RAP framework allows a SU-TX to probabilistically transmit over the sender-selected spectrum with a certain probability q (since not using the spectrum at all can lead to unnecessarily missing the opportunity). Otherwise, the SU-TX will search for another spectrum to use with probability $(1 - q)$. Here, the SU-TX only uses the minimum power/rate due to their robustness to interference and their weak impact on ongoing transmissions. The SU-TX does not gradually increase its rate and power any further as it still cannot exactly assess the its impact on the reception of nearby transmissions.

In Chap. 6, we calculate the optimal values of p and q that maximize the CRN goodput while satisfying the PRN performance guarantees. In what follows, we present the detailed protocol implementation of the RAP framework.

5.2 RAP-MAC Protocol

The rate-adaptive probabilistic medium access control (RAP-MAC) is the protocol implementation of the framework summarized in Algorithm 1 and described as follows.

SU-TX Spectrum Request (SR): Once a SU-TX has a packet in its transmit queue, it picks a frequency band at random (if there does not exist a preferred one that recently carried out a successful transmission). The secondary sender measures the interference in its vicinity and includes this measurement in a Spectrum Request

Algorithm 1 Pseudocode of the RAP-MAC protocol

```
SU-TX Spectrum Request
if current_spectrum = 0 then
    choose i ∈ {1,...,N} with probability 1/N
    current_spectrum = i
end if
P_int^tx = spectrum_measure(current_spectrum)
send(SR(current_spectrum, P_int^tx))

SU-RX Spectrum Grant
receive(SR(current_spectrum, P_int^tx))
P_int^rx = spectrum_measure(current_spectrum)
if (P_int^tx ≤ P_mask^(i)) and (P_int^rx ≤ P_mask^(i)) then
    clear_spectrum = 1
    send(SG(R_max^(i), clear_spectrum))
else if (P_int^tx > P_mask^(i)) and (P_int^rx ≤ P_mask^(i)) then
    clear_spectrum = 0
    send(SG(R_1, clear_spectrum))
end if

SU-TX Data Packet Transmission
receive(SG(r, clear_spectrum))
if clear_spectrum and Single_SU then
    rate = R_max^(i) with probability p
    rate = R_min with probability 1 − p
    send(DATA)
else if clear_spectrum and not Single_SU then
    rate = R_1
    send(DATA)
else
    rate = R_1
    send(DATA) with probability q
end if

SU-TX Receiving Acknowledgement
if receive(ACK) and R_min < R_(max−1)^(i) then
    Single_SU = 1
    increase(R_min)
else
    current_spectrum = 0
    Single_SU = 0
    R_min = R_1
end if
```

(SR) packet along with the identity of the selected spectrum. The SU-TX sends the SR packet over the common control channel to the destined receiver and sets the timeout timer.

SU-RX Spectrum Grant (SG): Upon receiving the SR packet, the receiver also measures the interference on the sender-selected spectrum. In clear spectrum scenarios, the receiver replies with a Spectrum Grant (SG) packet including the maximum rate, $R_{max}^{(i)}$, to be used over this spectrum along with an indicator that this spectrum is clear and has low interference levels. On the other hand, in unclear spectrum scenarios, the receiver includes the minimum allowable rate in the SG packet. In the remaining two cases of high interference at the receiver, the receiver will not reply with a SG packet as the transmission over the spectrum selected by the sender will not succeed given the high interference level in the receiver vicinity. After the sender timeout timer expires, the sender reselects another spectrum and resends a new SR to inform the receiver its new spectrum selection. Note that RAP-MAC uses the common control channel only to coordinate between a secondary sender and its respective receiver and not for inter-flow coordination as the case with the existing related literature [5–8]. In other words, the SR and SG packets are not used to deliver any information to neighboring cognitive radio nodes.

SU-TX Data Transmission: When the SU-TX receives the SG packet from its receiver within the timeout period, the SU-TX will transmit the data packet over the selected spectrum. However, it will only probabilistically use the rate selected by its receiver. According to RAP-MAC, if the SG packet indicates a clear spectrum, the sender will use the maximum permissible rate $R_{max}^{(i)}$ only with probability p and use the minimum rate $R_{min} = R_1$ with probability $1 - p$. As the transmission over the selected spectrum succeeds, the sender gradually increases R_{min} up to the second highest rate in order to minimize the capacity losses when there is no other nearby secondary transmissions. Meanwhile, if the SG packet indicates an unclear spectrum, the sender transmits over the selected spectrum only with probability q at the minimum rate R_1 or selects a new spectrum to sense with probability $1-q$. The SU-TX then transmits the data packet over the selected spectrum with the appropriate rate and sets the timeout timer.

SU-TX Receive Acknowledgement (ACK): If the secondary receiver successfully receives its intended data packet, it responds back with an ACK packet over the data channel. If the SU-TX receives the ACK packet before the timeout timer expires, it declares the used spectrum as its favorite spectrum for upcoming transmissions if the used rate is greater than R_1. Otherwise, the SU-TX sets its favorite spectrum to null. A packet is discarded if a maximum retry limit is reached.

References

1. Liu, H., Krishnamachari, B.: Randomized strategies for multi-user multi-channel opportunity sensing. In: Proceedings of IEEE CCNC Cognitive Radio Networks Workshop, Las Vegas, NV (2008)
2. Liang, Z., Liu, W., Zhou, P., Gao, F.: Randomized multi-user strategy for spectrum sharing in opportunistic spectrum access network. In: Proceedings of IEEE ICC Workshops, Beijing, China (2008)
3. Rappaport, T.: Wireless communications, principles & practice. Prentice Hall, Englewood Cliffs (1996)

4. Khattab, A.: The case for SIMO random access in multi-antenna multi-hop wireless networks. Wireless Networks (ACM/Springer) **17**(7), 1649–1665 (2011)
5. Zhao, Q., Tong, L., Swami, A., Chen, Y.: Decentralized cognitive MAC for opportunistic spectrum access in ad hoc networks: A POMPD framework. IEEE J. Sel. Area. Comm. **25**(3), 589–600 (2007)
6. Wang, F., Krunz, M., Cui, S.: Price-based spectrum management in cognitive radio networks. IEEE J. Sel. Top. Signal Process. **2**(1), 74–87 (2008)
7. Salameh, H.B., Krunz, M., Younis, O.: MAC protocol for opportunistic cognitive radio networks with soft guarantees. IEEE Trans. Mobile Comput. **8**(10), 1339–1352 (2009)
8. Huang, S., Liu, X., Ding, Z.: Optimal transmission strategies for dynamic spectrum access in cognitive radio networks. IEEE Trans. Mobile Comput. **8**(12), 1636–1648 (2009)

Chapter 6
RAP-MAC Performance Optimization with Statistical PRN Guarantees

Typically, the performance of a PRN is defined in terms of its outage probability [1–11]. For each primary user j in the ith PRN, the outage probability $p_{out}^{(i)}(PU_j)$ is bounded by β. The constrained CRN optimization problem is formulated as follows

$$\begin{aligned} &\text{maximize} \quad \sum_{i=1}^{N} \frac{1}{N} \cdot r_{SU}^{(i)} \\ &\text{subject to} \quad p_{out}^{(i)}(PU_j) \leq \beta \quad \forall i = 1,2,\ldots,N; j = 1,2,\ldots \end{aligned} \tag{6.1}$$

We next formulate this generic problem in terms of the RAP-MAC framework to find the optimal values of its parameters. For the ease of presentation, Table 6.1 lists the used notations.

6.1 RAP-MAC Achievable Flow Rate

First, we compute the average rate a SU can achieve over the ith band, $r_{SU}^{(i)}$, using the possible transmission rates and their corresponding RAP-MAC probabilities. Given the interference measurements at the sender and the receiver, there exists two possible cases that allow the secondary sender-receiver pair to use the randomly selected spectrum band. The first case is the clear spectrum case in which the interference measurements at both endpoints are below the power mask defines for this particular spectrum band. In the second case of unclear spectrum, only the interference measured at the secondary receiver is below the PRN power mask. Due to the independence of the interference measurements at the sender and its receiver, the probability of the two cases are $(\Pr[P_{int}^{(i)} \leq P_{mask}^{(i)}])^2$ and $\Pr[P_{int}^{(i)} \leq P_{mask}^{(i)}](1 - \Pr[P_{int}^{(i)} \leq P_{mask}^{(i)}])$, respectively, where P_{int} is the random variable representing the interference experienced at a SU terminal over the ith spectrum

A. Khattab et al., *Cognitive Radio Networks: From Theory to Practice*,
Analog Circuits and Signal Processing, DOI 10.1007/978-1-4614-4033-8_6,

Table 6.1 List of used notations

Parameter	Definition
n	Propagation path loss exponent
d_c	Distance beyond which the interference is negligible (i.e., below the receiver sensitivity)
$\lambda^{(i)}$	Operating wavelength of the ith PRN
$G_T^{(i)}$	Transmit antenna gain of the ith PRN
$G_R^{(i)}$	Receive antenna gain of the ith PRN
$d_o^{(i)}$	Close-in distance of the ith PRN
$P_o^{(i)}$	Reference power at the close-in distance of the ith PRN $P_o^{(i)} = \frac{P_{PU}^{(i)} G_T^{(i)} G_R^{(i)} \lambda^{(i)2}}{4\pi d_o^{(i)2}}$
α_i	Activity factor of the ith PRN
ρ_i	User density of the ith PRN
ρ_{SU}	User density of the CRN
$P_{mask}^{(i)}$	Power mask of the ith PRN
$P_{max}^{(i)}$	Maximum SU power to be used over the ith spectrum
$R_{max}^{(i)}$	Maximum SU rate to be used over the ith spectrum
$R_{max-1}^{(i)}$	Second highest SU rate to be used over the ith spectrum
R_1	Minimum SU rate
$\mathrm{erfc}(\cdot)$	Complementary error function [12]

band. The probability distribution of $P_{int}^{(i)}$ was approximated in [9] by a lognormal distribution with mean and variance given by

$$E[P_{int}^{(i)}] = \begin{cases} 2\pi\alpha_i\rho_i P_o^{(i)} d_o^{(i)^2} e^{-\pi\alpha_i\rho_i d_o^{(i)^2}} \ln \frac{d_c}{d_o^{(i)}}, & n = 2 \\ \frac{2\pi\alpha_i\rho_i P_o^{(i)} d_o^{(i)^2}}{n-2} e^{-\pi\alpha_i\rho_i d_o^{(i)^2}}, & n > 2 \end{cases} \tag{6.2}$$

and

$$Var[P_{int}^{(i)}] = \frac{\pi\alpha_i\rho_i}{n-1} \left[2P_o^{(i)} d_o^{(i)^2} e^{-\pi\alpha_i\rho_i d_o^{(i)^2}} \right]^2, \quad n \geq 2 \tag{6.3}$$

respectively, where n is the propagation path loss exponent of the environment, d_c is the distance beyond which the interference is negligible (i.e., below the receiver sensitivity), $d_o^{(i)}$ is the close-in distance and $P_o^{(i)} = \frac{P_{PU}^{(i)} G_T^{(i)} G_R^{(i)} \lambda^{(i)2}}{4\pi d_o^{(i)2}}$ is the reference power received at the close-in distance, $G_T^{(i)}$ is the operating wavelength of the ith PRN, $G_T^{(i)}$ and $G_R^{(i)}$ are the transmit and receive antenna gains, respectively.

Given the statistics of the distribution of $P_{int}^{(i)}$, the probabilities of the clear and unclear spectrum are given by

$$p_{clear} = \left[\frac{1}{2} \mathrm{erfc} \left(-\frac{\ln P_{mask}^{(i)} - \mu_{P_{int}^{(i)}}}{\sqrt{2\sigma^2_{P_{int}^{(i)}}}} \right) \right]^2 \tag{6.4}$$

and

$$p_{unclear} = \frac{1}{2}\mathrm{erfc}\left(-\frac{\ln P_{mask}^{(i)} - \mu_{P_{int}^{(i)}}}{\sqrt{2\sigma^2_{P_{int}^{(i)}}}}\right)\cdot\left[1-\frac{1}{2}\mathrm{erfc}\left(-\frac{\ln P_{mask}^{(i)} - \mu_{P_{int}^{(i)}}}{\sqrt{2\sigma^2_{P_{int}^{(i)}}}}\right)\right] \tag{6.5}$$

respectively, where erfc$(\cdot)$ is the complementary error function [12], and

$$\mu_{P_{int}^{(i)}} = \ln(E[P_{int}^{(i)}]) - \frac{1}{2}\ln\left(1+\frac{Var[P_{int}^{(i)}]}{E[P_{int}^{(i)}]^2}\right) \tag{6.6}$$

$$\sigma^2_{P_{int}^{(i)}} = \ln\left(1+\frac{Var[P_{int}^{(i)}]}{E[P_{int}^{(i)}]^2}\right) \tag{6.7}$$

According to RAP-MAC, the rate of a sender-receiver pair is qR_1 in the unclear spectral opportunity case. We next calculate the average secondary flow rate whenever the spectrum is measured to be clear. The flow rate given no other secondary senders are in the vicinity of the tagged secondary receiver and using the selected band is $pR_{max}^{(i)} + (1-p)R_{max-1}^{(i)}$, where $R_{max}^{(i)}$ is the maximum usable rate over this spectrum.[1] Meanwhile, the flow rate is R_1 if there exists at least one more SU transmitting on the selected spectrum in the vicinity of the tagged secondary receiver. The probability of having at least one more secondary sender over the selected band in the receiver's vicinity is the probability of having $k \geq 2$ secondary senders and one minus the probability of only the tagged sender selecting the ith band while the remaining $k-1$ senders select different bands. Since the locations of the secondary users are modeled as a homogeneous Poisson process, the probability of the number of potential senders within a disc area $A_c = \pi d_c^2$ is equal to k is given by

$$\Pr[K=k] = \frac{e^{-\rho_{SU}A_c}(\rho_{SU}A_c)^k}{k!}, \quad k=0,1,2,\ldots \tag{6.8}$$

Hence, the probability of multiple concurrent secondary transmissions over the ith band, $p_{multiuser}$, is given by

$$\begin{aligned} p_{multiuser} &= \sum_{k=2}^{\infty}\frac{e^{-\rho_{SU}A_c}(\rho_{SU}A_c)^k}{k!}\cdot\left(1-\frac{1}{N}\left(\frac{N-1}{N}\right)^{k-1}\right) \\ &= 1-e^{-\rho_{SU}A_c} - \frac{e^{-\frac{\rho_{SU}A_c}{N}}}{N-1} + \frac{e^{-\rho_{SU}A_c}}{N-1} \end{aligned} \tag{6.9}$$

[1] We do not incorporate the ramp up from R_1 to R_{max-1}. While such assumption slightly impact the achievable rate of a SU, it does not affect our optimization problem as the outage constraints depend only on the maximum used rate.

where $(1-\frac{1}{N}(\frac{N-1}{N})^{k-1})$ is the probability that at least one other SU sender selects the same band. Similarly, the probability of no other concurrent secondary transmission, $p_{singleuser}$ is computed using the probability of the two events of either no other nearby sender exists (i.e., the probability of $k<2$) or non of the $k\geq 2$ nearby senders selects the same band as the tagged sender as

$$\begin{aligned} p_{singleuser} &= e^{-\rho_{SU}A_c}(1+\rho_{SU}A_c) \\ &\quad + \sum_{k=2}^{\infty} \frac{e^{-\rho_{SU}A_c}(\rho_{SU}A_c)^k}{k!} \cdot \frac{1}{N}\left(\frac{N-1}{N}\right)^{k-1} \\ &= e^{-\rho_{SU}A_c} + \frac{e^{-\frac{\rho_{SU}A_c}{N}}}{N-1} - \frac{e^{-\rho_{SU}A_c}}{N-1} \end{aligned} \tag{6.10}$$

Using the probabilities of clear and unclear spectrum given by (6.4) and (6.5) and the multiple and single SU probabilities given by (6.9) and (6.10), the average rate of a SU is expressed as

$$r_{SU}^{(i)} = \left[\left(pR_{max}^{(i)} + (1-p)R_{max-1}^{(i)}\right) p_{singleuser} + R_1 p_{multiuser}\right] p_{clear} + qR_1 p_{unclear} \tag{6.11}$$

6.2 Statistical PRN Outage Constraints

Next, we formally define the statistical constraints on the outage probability given in (6.1) in terms of p, q, and the maximum secondary user transmission power over different spectrum bands. For a given secondary transmitter, all of the surrounding primary receivers must successfully receive their intended data with probability $1-\beta$. This constraint is satisfied if and only if it is satisfied at the primary receiver that is closely located with respect to the secondary sender. Let's denote the minimum distance between a secondary sender and the closest primary receiver by $\mathscr{D}_{min}$. We define the outage probability $p_{out}^{(i)}$ at the ith PRN receiver at distance $\mathscr{D}$ as follows

$$\begin{aligned} p_{out}^{(i)} &= \Pr[\text{SU-TX}]\left(\Pr\left[\text{outage}|\mathscr{D} < \mathscr{D}_{min}^{(i)}\right]\Pr\left[\mathscr{D} < \mathscr{D}_{min}^{(i)}\right]\right. \\ &\quad \left. + \Pr\left[\text{outage}|\mathscr{D} \geq \mathscr{D}_{min}^{(i)}\right]\Pr\left[\mathscr{D} \geq \mathscr{D}_{min}^{(i)}\right]\right) \end{aligned} \tag{6.12}$$

where Pr[SU-TX] is either p, or q depending on the interference measurements at the secondary flow endpoints, and $\mathscr{D}_{min}^{(i)}$ is a random variable that models the minimum distance between a secondary sender and a primary receiver in the ith PRN. The probabilities of the two events $\mathscr{D} < \mathscr{D}_{min}^{(i)}$ and $\mathscr{D} \geq \mathscr{D}_{min}^{(i)}$ are computed using the cumulative distribution of the minimum distance between a SU-TX and a PU-RX studied in [9, 13]. According to our system model, the cumulative distribution function of $\mathscr{D}_{min}^{(i)}$ is given by

$$F_{\mathscr{D}_{min}^{(i)}}(d) = \Pr[\mathscr{D}_{min}^{(i)} < d] = 1 - e^{-\pi\alpha_i\rho_i d^2} \tag{6.13}$$

Let's define $\mathscr{D}_{min}^{(i)*}$ to be the minimum distance below which the probability of outage is unity, that is, $\Pr[\text{outage}|\mathscr{D} < \mathscr{D}_{min}^{(i)*}] \triangleq 1$. According to (6.13), $\mathscr{D}_{min}^{(i)}$ is at least $\mathscr{D}_{min}^{(i)*}$ with probability $p_{\mathscr{D}_{min}^*} = 1 - \Pr[\mathscr{D}_{min} < \mathscr{D}_{min}^{(i)*}]$. Substituting in (6.13), we get

$$\mathscr{D}_{min}^{(i)*} = \sqrt{\frac{-\ln(p_{\mathscr{D}_{min}^*})}{\pi\alpha_i\rho_i}} \tag{6.14}$$

Note that $p_{\mathscr{D}_{min}^*}$ determines how much $\mathscr{D}_{min}^{(i)}$ is close to $\mathscr{D}_{min}^{(i)*}$. Given that $\Pr[\text{outage}|\mathscr{D} < \mathscr{D}_{min}^{(i)*}] \triangleq 1$, and let $\gamma^{(i)}$ denotes the conditional outage probability $\Pr[\text{outage}|\mathscr{D} \geq \mathscr{D}_{min}^{(i)*}]$, the outage probability given by (6.12) can be rewritten as

$$p_{out}^{(i)} = \Pr[\text{SU-TX}]\left(\left(1 - p_{\mathscr{D}_{min}^*}\right) + \gamma^{(i)} p_{\mathscr{D}_{min}^*}\right) \tag{6.15}$$

Hence, the $p_{out}^{(i)} \leq \beta$ constraints in (6.1) are equivalent to

$$\gamma^{(i)} \leq 1 - \frac{1 - \frac{\beta}{\Pr[\text{SU-TX}]}}{p_{\mathscr{D}_{min}^*}} \tag{6.16}$$

Since $\gamma^{(i)}$ cannot be negative, Pr[SU-TX] must be no less than β and the following constraint must be satisfied

$$\Pr[\text{SU-TX}] \leq \frac{\beta}{1 - p_{\mathscr{D}_{min}^*}} \tag{6.17}$$

Finally, we relate the outage probability of the ith band to the ith PRN power mask and the maximum power a SU can use over that band. In order to preserve the required bounds on $p_{out}^{(i)}(PU_j)$, the following condition at every primary receiver j should be satisfied with probability $(1-\gamma^{(i)})\Pr[\text{SU-TX}]$ due to every secondary transmission

$$P_{int,j}^{(i)} + g_{\mathscr{D}_{min}^*}^{(i)} P_{SU}^{(i)} \leq P_{mask}^{(i)} \tag{6.18}$$

where $P_{int,j}^{(i)}$ is the interference power at the jth primary receiver due to other potential interfering activities, and $g_{\mathscr{D}_{min}^*}^{(i)} = \frac{G_T^{(i)} G_R^{(i)} \lambda^{(i)2}}{(4\pi)^2(\mathscr{D}_{min}^{(i)*})^n}$ is the channel gain between the nearest secondary sender and the jth primary receiver. Since RAP-MAC allows a secondary sender to use different transmission powers with certain probabilities, it is sufficient that the maximum permissible power $P^{(i)}_{max}$ which is used with probability Pr[SU-TX] $= p$ satisfies the condition in (6.18).[2] In order to satisfy (6.18) with probability $(1-\gamma^{(i)})p$, we compute the $[(1-\gamma)p]$-quantile of $P^{(i)}_{int,j}$ and substitute in (6.18). According to [9], $P^{(i)}_{int,j}$ has a lognormal distribution, and hence, its $[(1-\gamma^{(i)})p]$-quantile $P^{(i)}_{(1-\gamma)p}$ is calculated as

$$P^{(i)}_{(1-\gamma)p} = \exp(-\sqrt{2Var[P^{(i)}_{int}]}\mathrm{erfc}^{-1}(2(1-\gamma^{(i)})p)) \tag{6.19}$$

Substituting with (6.19) in (6.18), we get the following constraint on the maximum transmission power of a secondary user over the ith band

$$P^{(i)}_{max} \leq \frac{P^{(i)}_{mask} - P^{(i)}_{(1-\gamma)p}}{g^{(i)}_{\mathscr{D}^*_{min}}} \tag{6.20}$$

6.3 RAP-MAC Parameter Optimization

Given $r^{(i)}_{SU}$ formulated in terms of p and q as in (6.11), the general optimization problem given in (6.1) can be restated in terms of the RAP-MAC parameters as follows

$$\begin{aligned} \text{maximize} \quad & \sum_{i=1}^{N} \frac{1}{N} \cdot r^{(i)}_{SU} \\ \text{subject to} \quad & P^{(i)}_{max} \leq \frac{P^{(i)}_{mask} - P^{(i)}_{(1-\gamma)p}}{g^{(i)}_{\mathscr{D}^*_{min}}} \quad \forall i = 1,2,\ldots,N \\ & \beta \leq p \leq \frac{\beta}{1-p_{\mathscr{D}^*_{min}}} \\ & \beta \leq q \leq \frac{\beta}{1-p_{\mathscr{D}^*_{min}}} \end{aligned} \tag{6.21}$$

[2]In multiuser scenarios, RAP-MAC uses lower powers/rates. The interference caused by multiple weak sources has negligible impact (almost as Additive White Gaussian Noise (AWGN)) on ongoing transmissions compared to the interference from a single high power/rate source [12, 14]. Hence, (6.19) also covers the multiuser case.

This is a mixed integer non-linear programming problem the solution of which is the optimal values of p and q as well as the maximum permissible SU transmit powers $P_{max}^{(i)}$ (and hence, the corresponding maximum transmission rates $R_{max}^{(i)}$) over each of the N bands. Solving such a mixed integer non-linear programming problem is NP hard. In what follows, we present an exhaustive study of the impact of different factors over the solution of the problem and hence the achievable CRN user rate. We use the MATLAB optimization toolbox [15] for our study of the solution of the RAP-MAC problem. We consider 4 PRNs distributed over a $500 \times 500\,\text{m}^2$ area each with 200 users using the $\{0.769, 0.925, 2.412, 5.180\}$ GHz channels with power masks of 2 nW and channel bandwidth $B_i = 20\,\text{MHz}$ for all channels. Other simulation parameters are $d_o = \{42, 33, 12, 6\}$ cm, $P_{PU}^{(i)} = 1$ W, $G_T^{(i)} = G_R^{(i)} = 1$ for all i, $n = 4$, and $d_c = 50\,\text{m}$ for $-80\,\text{dB}$ receive sensitivity. A SU-TX picks its rate from $\{54, 36, 24, 12, 2\}$ Mbps with the power of the 54 Mbps rate is 1 W and the corresponding power of other rates is computed using (4.1).

6.3.1 Impact of $\mathbf{p}_{\mathscr{D}^*_{\mathbf{min}}}$

The only variable in the above problem formulation is $p_{\mathscr{D}^*_{min}}$ which reflects the accuracy of the minimum distance between a secondary sender and a primary receiver. Figure 6.1 depicts the optimal p and q values and the CRN user rate versus the PRN activity factor for different $p_{\mathscr{D}^*_{min}}$ values for $\beta = 5\%$. As shown in Fig. 6.1a, the optimal probability of transmission over a clear spectral opportunity, p, depends significantly on the choice of $p_{\mathscr{D}^*_{min}}$ and tends to be the maximum possible value of $\beta/(1 - p_{\mathscr{D}^*_{min}})$. However, the PRN activity factor does not impact p as p is the probability of using the highest possible power/rate conditioning on the lack of nearby PRN activities. On the other hand, q, the probability of SU transmission given PRN activities in the vicinity of the SU-TX, varies with both $p_{\mathscr{D}^*_{min}}$ and the PRN activity factor as illustrated in Fig. 6.1b. As the PRN activities increase, q also increases to allow RAP-MAC to explore potentially missed opportunities more frequently to maximize the CRN user rate.

6.3.2 Impact of the PRN Outage Constraint

Next, we evaluate the impact of the maximum outage probability allowed by the PRNs, β. We solve (6.21) for β equals to 1, 5, and 10%. For the stringent outage constraint of $\beta = 1\%$, both p and q fall rapidly as $p_{\mathscr{D}^*_{min}}$ decreases as shown in Fig. 6.2. Recall that $p_{\mathscr{D}^*_{min}}$ represents how $\mathscr{D}^*_{min}$ is close to the distance at which outage occurs with probability equal to unity. Hence, as $p_{\mathscr{D}^*_{min}}$ decreases RAP-MAC tends to be more conservative (i.e., lower p and q values) in order not to violate the

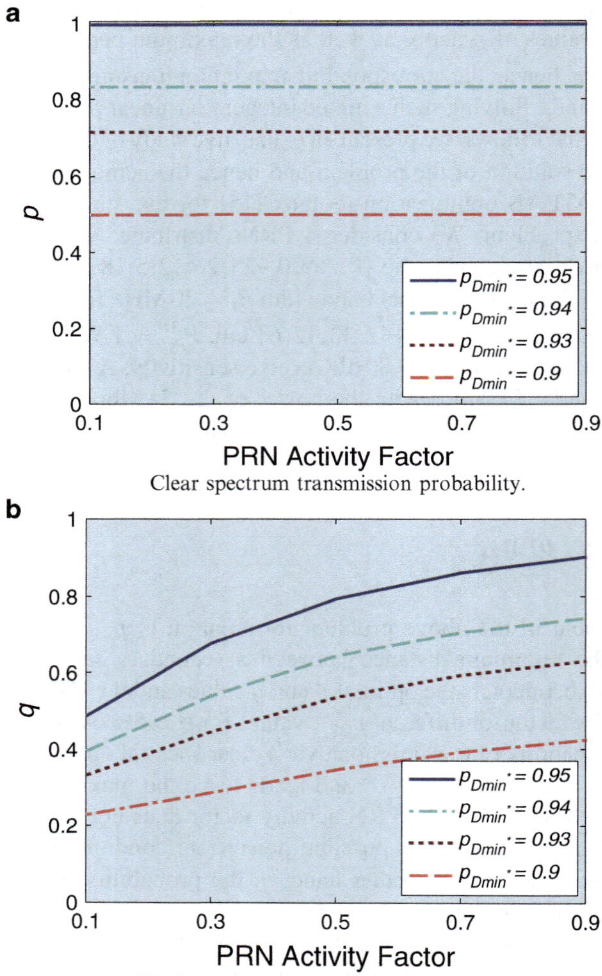

(a) Clear spectrum transmission probability.

(b) Unclear spectrum transmission probability.

Fig. 6.1 Optimal transmission probabilities for different PRN activity factors and $p_{\mathscr{D}^*_{min}}$ values for $\beta = 5\%$

PRN constraints. However, as β increases, the impact of $p_{\mathscr{D}^*_{min}}$ on the optimal values of p and q is reduced. As shown in Fig. 6.2 for $\beta = 5\%$, the values of p and q fall slower with the decrease of the value of $p_{\mathscr{D}^*_{min}}$ and even more slower for $\beta = 10\%$. Note that the PRN activity factor only impacts the value of q (but not p) as explained earlier regardless of the value of β. However, the impact of the PRN activity factor on q increases with the relaxation of the PRN constraint β as shown in Fig. 6.2b.

Fig. 6.2 Impact of β and $p_{\mathscr{D}^*_{min}}$ on the optimal transmission probabilities for different PRN activity factors

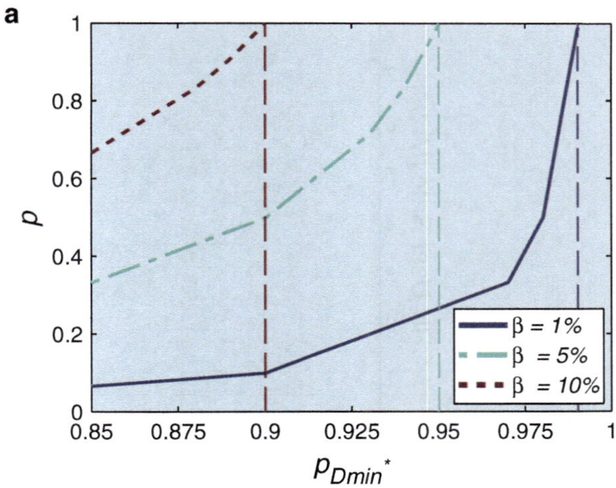

Clear spectrum transmission probability.

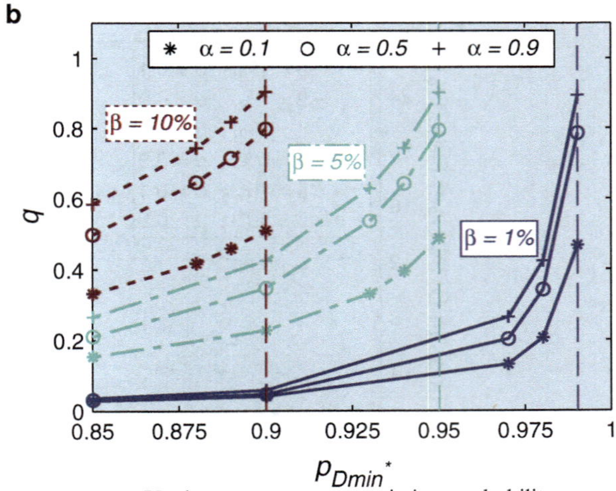

Unclear spectrum transmission probability.

6.3.3 *CRN User Rate*

Despite the strong dependence of the optimal p and q values on $p_{\mathscr{D}^*_{min}}$, Fig. 6.3a shows that $p_{\mathscr{D}^*_{min}}$ has a minimal impact on the maximum rate of CRN users. While the closer $p_{\mathscr{D}^*_{min}}$ to $1-\beta$ achieves the highest CRN rate, using smaller values for $p_{\mathscr{D}^*_{min}}$ achieves very close CRN rates. For, example, the CRN user rate using $p_{\mathscr{D}^*_{min}} = 0.94$ is only 1–2.8% (depending on the PRN activity factor) less than the rate when $p_{\mathscr{D}^*_{min}} = 0.95$. Recall that the CRN user rate deteriorates with the increase of the PRN activity. Meanwhile, using $p_{\mathscr{D}^*_{min}} = 0.94$ in stead of 0.95 changes p from 0.833

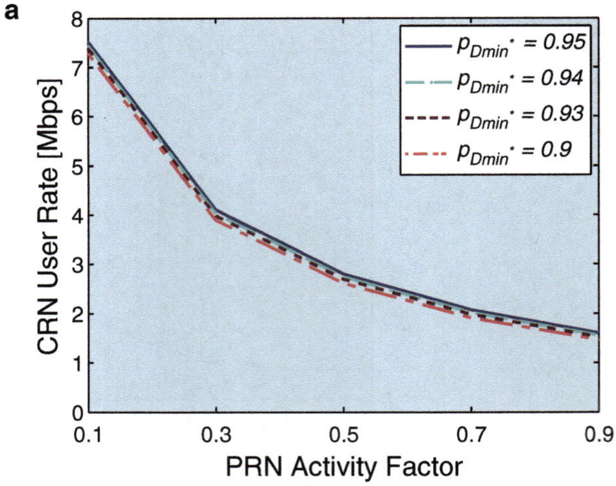

CNR flow rate for β = 5%.

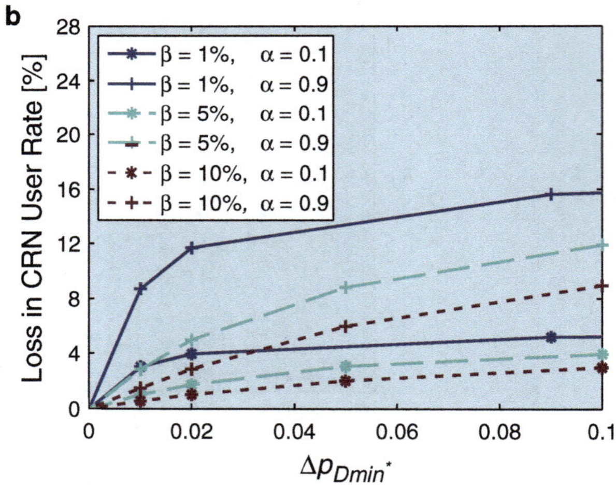

Loss is CNR flow rate versus the offset in $p_{\mathscr{D}^*_{min}}$.

Fig. 6.3 The optimal CRN user rate and the impact of β, α, and $p_{\mathscr{D}^*_{min}}$

to 0.714 which allows a bigger probabilistic capacity margin for multiple SUs to share the available opportunities. Similar results were obtained for other values of β. Figure 6.3b depicts the loss in the CRN user rate versus the offset in $p_{\mathscr{D}^*_{min}}$ from its maximum value of $1-\beta$ for different values of β and α. The deterioration in the CRN user rate with $p_{\mathscr{D}^*_{min}}$ increases as the PRN constraint β gets tighter and the PRN activity factor α increases.

References

1. Akyildiz, I.F., Lee, W.Y., Chowdhury, K.R.: CRAHNs: Cognitive radio ad hoc networks. Ad Hoc Networks (Elsevier) **7**(5), 810–836 (2009)
2. Salameh, H.B., Krunz, M.: Channel access protocols for multihop opportunistic networks: challenges and recent developments. IEEE Networks **23**(4), 14–19 (2009)
3. Yucek, T., Arslan, H.: A survey of spectrum sensing algorithms for cognitive radio applications. IEEE Comm. Surv. Tutorials **11**(1), 116–130 (2009)
4. Tian, Z., Giannakis, G.: Compressed sensing for wideband cognitive radios. In: Proceedings of IEEE ICASSP, Honolulu, HI (2007)
5. Liu, H., Krishnamachari, B.: Randomized strategies for multi-user multi-channel opportunity sensing. In: Proceedings of IEEE CCNC Cognitive Radio Networks Workshop, Las Vegas, NV (2008)
6. Liang, Z., Liu, W., Zhou, P., Gao, F.: Randomized multi-user strategy for spectrum sharing in opportunistic spectrum access network. In: Proceedings of IEEE ICC Workshops, Beijing, China (2008)
7. Zhao, Q., Tong, L., Swami, A., Chen, Y.: Decentralized cognitive MAC for opportunistic spectrum access in ad hoc networks: A POMPD framework. IEEE J. Sel. Area. Comm. **25**(3), 589–600 (2007)
8. Wang, F., Krunz, M., Cui, S.: Price-based spectrum management in cognitive radio networks. IEEE J. Sel. Top. Signal Process. **2**(1), 74–87 (2008)
9. Salameh, H.B., Krunz, M., Younis, O.: MAC protocol for opportunistic cognitive radio networks with soft guarantees. IEEE Trans. Mobile Comput. **8**(10), 1339–1352 (2009)
10. Anandkumar, A., Michael, N., Tang, A.: Opportunistic spectrum access with multiple users: Learning under competition. In: Proceedings of IEEE INFOCOM 2010, San Deigo, CA (2010)
11. Chaporkar, P., Proutiere, A., Asnani, H.: Learning to optimally exploit multi-channel diversity in wireless systems. In: Proceedings of IEEE INFOCOM 2010, San Diego, CA (2010)
12. Rappaport, T.: Wireless communications, principles & practice. Prentice Hall, Englewood Cliffs (1996)
13. Clancy, T.C.: Achievable capacity under the interference temperature model. In: Proceedings of IEEE INFOCOM, Anchorage, AK (2007)
14. Khattab, A.: The case for SIMO random access in multi-antenna multi-hop wireless networks. Wireless Networks (ACM/Springer) **17**(7), 1649–1665 (2011)
15. MATLAB: version 7.10.0 (R2010a). The MathWorks Inc., Natick (2010)

Chapter 7
RAP-MAC Performance Evaluation in Large-Scale Networks

7.1 Simulations Setup

We develop an event-driven packet-level simulator using MATLAB [1]. We consider 9 PRNs collocated with a CRN in a $500 \times 500\,\mathrm{m}^2$ area. Each network has 200 nodes forming 100 sender-receiver pairs. The operating frequencies of the 9 PRNs are $\{0.769, 0.789, 0.809, 2.412, 2.432, 2.462, 5.180, 5.200, 5.220\}$ GHz with respective activity factors of $\{0.1, 0.5, 0.9, 0.1, 0.5, 0.9, 0.1, 0.5, 0.9\}$. The bandwidth of each channel is 20 MHz and the power mask is 2 nW for all PRNs. The PRN transmit power is 1 W and the transmit and receive antenna gains are equal to unity for all PRNs. We consider PRN maximum allowed outage probability values of 1, 5, and 10%. The path loss exponent n is set to be 4. A secondary transmission can use a rate in the set {54, 36, 24, 12, 2} Mbps. The corresponding set of transmission powers is calculated according to (4.1) with the transmission power of the 54 Mbps rate is equal to 1 W. We vary the arrival rate of all CRN users from 1 to 35 Mbps. For each arrival rate value, we generate ten random node topologies. For each topology, we generate three traffic matrices. The reported results are the average of these 30 runs for each arrival rate value. The error bars represent the 95% confidence interval of the multiple runs. We use (6.21) to compute the optimal values of p and q for different values of β.

Our benchmark is a protocol that belongs to the family of hypothetically-optimal spectrum access protocols which has a wide-band sensing capability and a greedy spectrum access approach in the sense that a SU-TX exploits the best spectral opportunity at the maximum permissible power/rate. We use [2] to compute such maximum powers/rates. In order to insure fairness in comparison, we do not implement the capability of a secondary user to simultaneously transmit over multiple spectrum bands at a given time instant as in the protocol presented in [2]. We refer to such a modified protocol as OPT-MAC as it represents a wide range of spectrum access protocols that adopt greedy spectrum access mechanisms for transmission over available spectral opportunities (e.g., [3–5]). OPT-MAC spectrum access mechanism is based on carrier-sensing and uses message exchange over

A. Khattab et al., *Cognitive Radio Networks: From Theory to Practice*,
Analog Circuits and Signal Processing, DOI 10.1007/978-1-4614-4033-8_7,

the common control channel to insure a single secondary user transmission per contention area. For each randomly generated topology and arrival process, we run both the RAP-MAC and OPT-MAC protocols to guarantee fairness in comparison. Data packets are 1,500 bytes long for both protocols. Control packets of both protocols are 40 bytes transmitted at 12 Mbps rate over the common control channel. Spectrum sensing and transceiver turn-around times are 9 and 5 μs, respectively. The exponential backoff window of OPT-MAC is bounded by (16, 1024) slots of duration 2 μs. Our performance metrics are the CRN average goodput, Jain's index as a measure of the fairness in the CRN goodput distribution [6], and the outage probability of the PRNs defined as the probability of PRNs transmission failure due to the CRN activities.

7.2 CRN Goodput Performance

Figure 7.1a depicts the average goodput of the CRN users using both the RAP-MAC and OPT-MAC for β equals to 5%. RAP-MAC achieves *significantly* higher goodput compared to OPT-MAC. The RAP-MAC gain in the CRN user goodput varies between 65 and 119.5% depending on the CRN traffic demand. RAP-MAC significant goodput gain is attributed to: (1) the fact that RAP-MAC probabilistically (with probability q) explores the spectral opportunities that OPT-MAC does not explore *at all* when the interference measurements imply unclear opportunities. (2) RAP-MAC is less susceptible to transmission failures (compared to OPT-MAC) due to its probabilistic and non-greedy policy in the clear spectrum situations which allows multiple flows to simultaneously use a given spectrum at the more robust lower transmission rates. Consequently, the RAP-MAC gain depends on the value of β which affects the optimal values of p and q as explained in Chap. 6. Figure 7.1b depicts the RAP-MAC gain for different β values. As β increases, the gain in the CRN goodput increases up to 138% as the case with β equals to 10%. Note that the value of q obtained using (6.21) is 0.5, 0.41, and 0.15 for β equals to 10%, 5%, and 1%, respectively. Furthermore, Fig. 7.1b shows that the gain peaks at low CRN traffic demands then decreases before it linearly increases with the traffic demands at β equals to 5 and 10%. For instance, maximum gains of 119.5% and 138% are achieved at 5 Mbps and 7.5 Mbps for β equals to 5% and 10%, respectively. However, the RAP-MAC goodput gain first decreases before increasing for the more stringent outage constraint of β equals to 1%.

The superior goodput performance of RAP-MAC is due to the bigger gap between its transmission attempts and transmission blockages (due to either PRN or CRN activities) compared to OPT-MAC as shown in Fig. 7.2a for β equals to 5%. As β increases, the gap between the blocked and attempted transmissions increases. Regardless the value of β, the number of transmission attempts of RAP-MAC (the solid stared curve) is only slightly higher than that of OPT-MAC (the solid circled curve). However, OPT-MAC transmissions are susceptible to more blockages (the dashed curves) as OPT-MAC does not account for the activities of

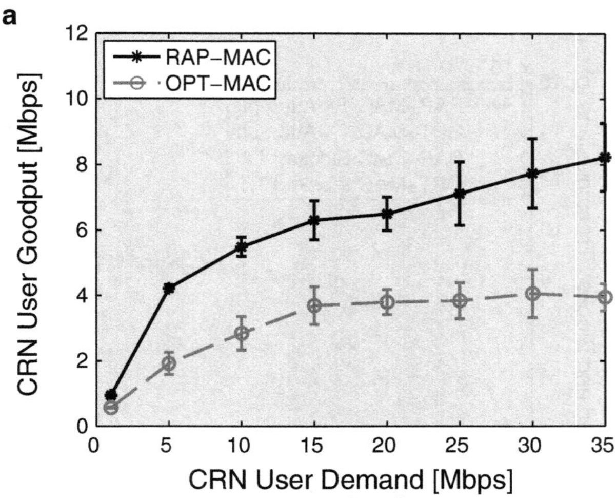

CRN user goodput for $\beta = 5\%$.

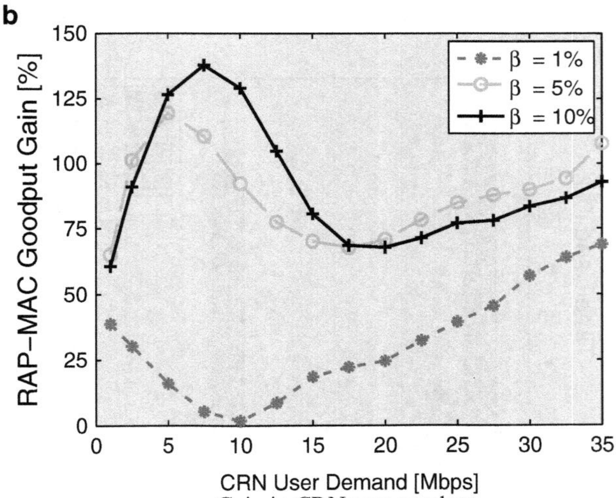

Gain in CRN user goodput.

Fig. 7.1 RAP-MAC achieves significantly higher goodput for different values of β

hidden PRN or CRN nodes. Recall that OPT-MAC either allows a CRN sender to transmit at the highest possible power/rate or to not transmit at all for a given spectrum band. Meanwhile, RAP-MAC has a secondary flow probabilistically adapt its power/rate based on the interference scenario. Figure 7.2b depicts the distribution of the rates used by RAP-MAC under low and high CRN traffic demands. At high CRN demands, RAP-MAC tends to have the CRN flows using the minimum rate more often to allow multiple CRN flows to simultaneously share the spectral opportunities. As the CRN demands decrease, RAP-MAC tends to use higher

a

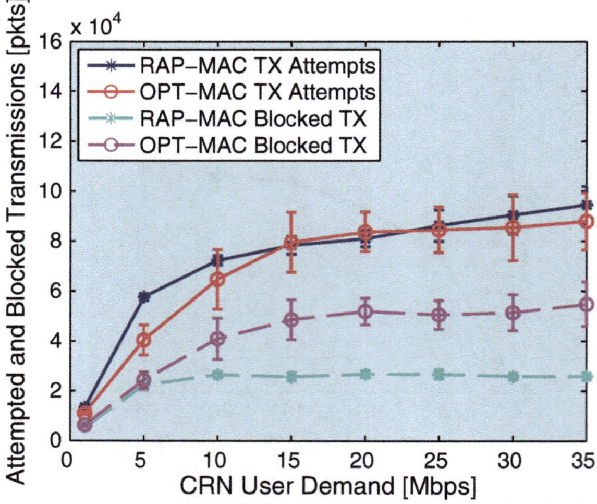

Attempted and blocked transmission attempts for $\beta = 5\%$.

b

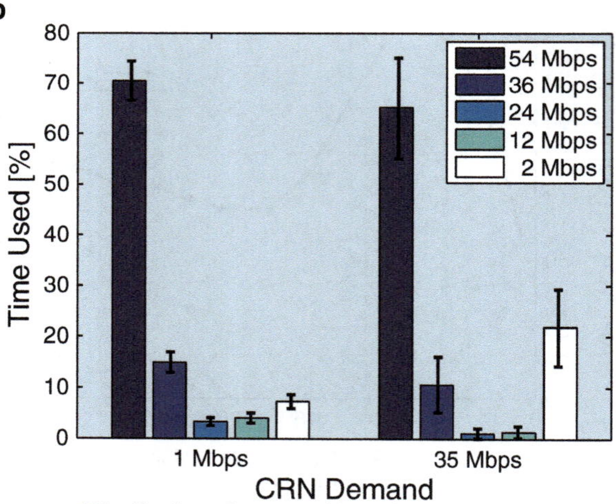

Distribution of rate used by RAP-MAC for $\beta = 5\%$.

Fig. 7.2 RAP-MAC spectrum access decisions lead to fewer blocked transmission attempts as it allows more low rate/power flows to coexist

rates (combined with switching to a different spectrum in some cases) in order to minimize the unutilized capacity of the available spectral opportunities. As shown in Fig. 7.2b, RAP-MAC uses R_{min} for only 7.3% of the time at 1 Mbps CRN demand compared to 22% at 35 Mbps CRN demand. Figure 7.2 explains how RAP-MAC spectrum access decisions result in higher goodput as was illustrated in Fig. 7.1.

7.3 CRN Fairness Performance

RAP-MAC does not have an explicit mechanism for inter-flow coordination. However, it adopts a probabilistic non-greedy transmission approach that prevents a single CRN flow from exclusively capturing an available spectral opportunity. This results in RAP-MAC significantly outperforming OPT-MAC in terms of the fairness characteristics as shown in Fig. 7.3. Figure 7.3a depicts Jain's Fairness

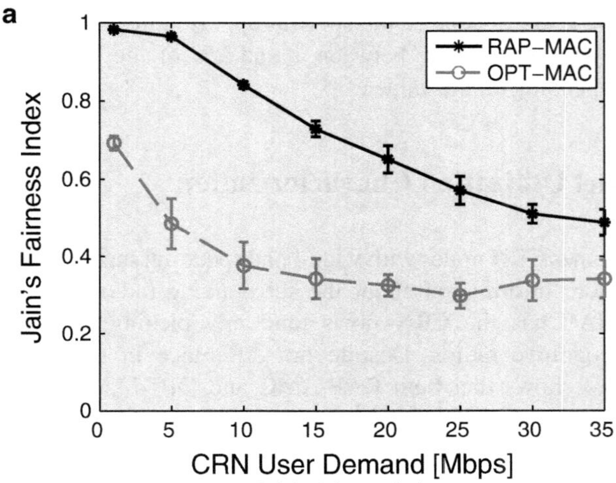

Jain's fairness index.

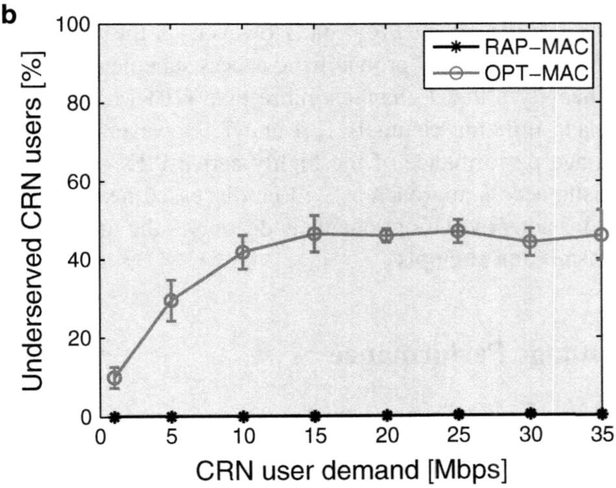

Percentage of underserved CRN flows.

Fig. 7.3 RAP-MAC fairly distributes the spectral opportunities among different users without explicit inter-flow coordination

Index defined as JFI $= \frac{(\sum_i T(i))^2}{L \sum_i T(i)^2}$, where $T(i)$ is the goodput of the ith flow and L is the number of CRN flows [6]. At low CRN demands, JFI of RAP-MAC approaches its optimal value of unity implying that all flows are getting approximately equal goodput shares. As the traffic demands increase, JFI of RAP-MAC decreases but it is always much higher than JFI of OPT-MAC. The poor fairness performance of OPT-MAC is attributed to its greedy transmission strategy that can allow some flows to exclusively capture spectral opportunities and thereby starving other flows in distributed ad-hoc networks. Figure 7.3b illustrates the percentage of flows receiving less than 10% of the average CRN goodput. It can be seen that OPT-MAC allows only 53% of the flows to capture the spectral opportunities starving the remaining 47% of the flows. Meanwhile, between 1 and 2% of the RAP-MAC flows are underserved depending on the value of β.

7.4 Channel Utilization Characterization

OPT-MAC assumes CRN nodes with wide-band spectrum sensing capability aiming at gathering more information about the surrounding radio environment. Meanwhile, RAP-MAC has the CRN flows randomly picking their channels given narrow-band cognitive radios. Despite the difference in the spectrum sensing scheme, Fig. 7.4 shows that both RAP-AMC and OPT-MAC tend to utilize the channels licensed to the PRNs with the lowest activity factors of 0.1 for most of the time, namely, channels 1, 4 and 7 illustrated by the dark blue, light blue and orange bars, respectively. At low CRN traffic demands, both RAP-MAC and OPT-MAC do not frequently utilize the rest of the channels with activity factors of 0.5 and 0.9 as illustrated in Fig. 7.4a. However, as the CRN traffic demands increase (Fig. 7.4b), RAP-MAC probabilistic access scheme allows the CRN flows to explore the heavily utilized channels more than OPT-MAC rather than having the excess demand utilizing channels 1, 4 and 7. However, RAP-MAC does not degrade the outage performance of the highly active PRNs because of the RAP-MAC probabilistic access approach as will be discussed next. Such a distribution of packet transmissions over more channels decreases the amount of blocked and failed CRN transmission attempts.

7.5 PRN Outage Performance

Finally, we investigate the outage performance of the primary licensed networks. Figures 7.5, 7.6, and 7.7 depict the outage probability of PRNs using channel 1, 4 and 7, respectively, for both RAP-MAC and OPT-MAC for β equals to 5%. These channels are highly exploited by the CRN according to both RAP-MAC and OPT-MAC. Recall that the frequency of channel 1 is lower than that of channel 4 which is lower than the frequency of channel 7. Due to the better propagation characteristics at lower frequencies, both the RAP-MAC and the OPT-MAC protocols favor

Fig. 7.4 The distribution of channel utilization for β equals to 5%

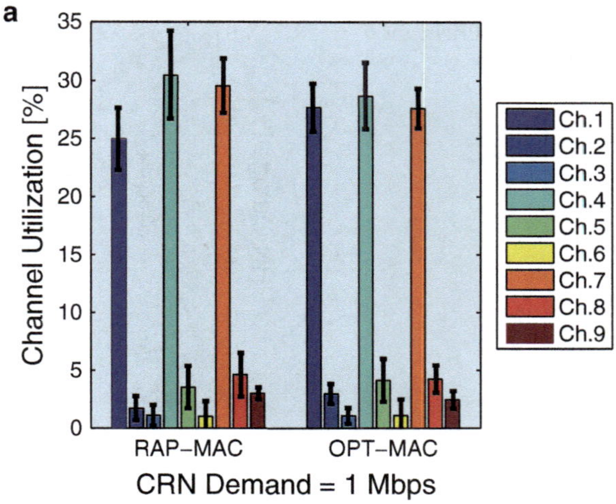

Low CRN demand.

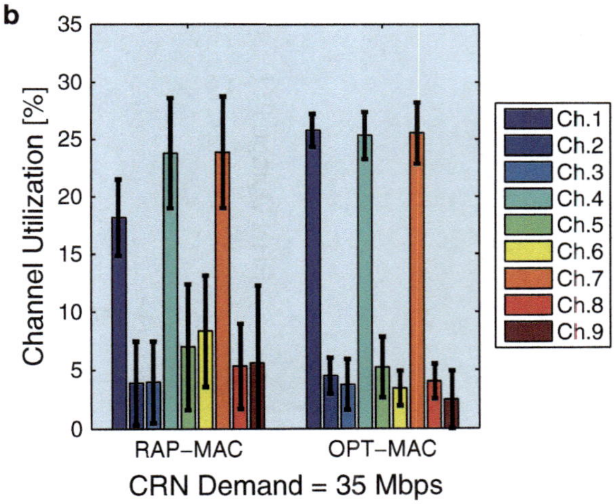

High CRN demand.

(i.e., initiate more transmission attempts) channel 1 more than channel 4 more than channel 7. Consequently, the outage probability of the PRN using channel 1 is the highest and the that of channel 7 is the lowest. For other PRNs with higher activity factors, the outage probability is insignificant for both protocols (below 0.001 and hence are not shown here). For all channels, the PRN outage probability due to RAP-MAC (represented by solid curves) is higher than that due to OPT-MAC (represented by dashed curves) because of the RAP-MAC probabilistic transmission policy and random sensing mechanism. However, the outage due to RAP-MAC is always below the PRN specified bound irrespective of the value of β and the CRN traffic demand. The superior RAP-MAC goodput gain outweighs such an excess PRN outages.

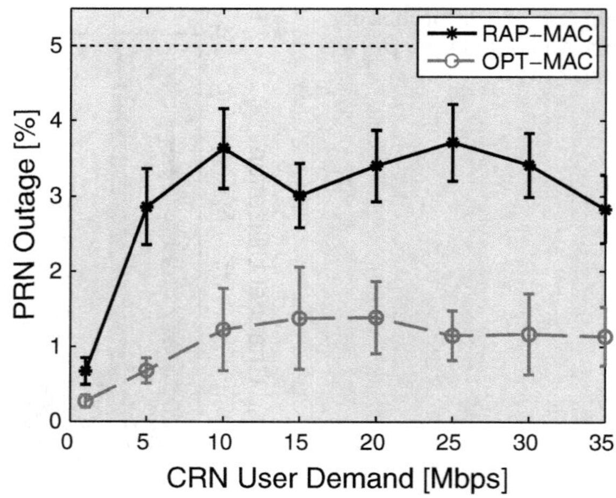

Fig. 7.5 PRN outage probability of PRN 1 for β equals to 5%

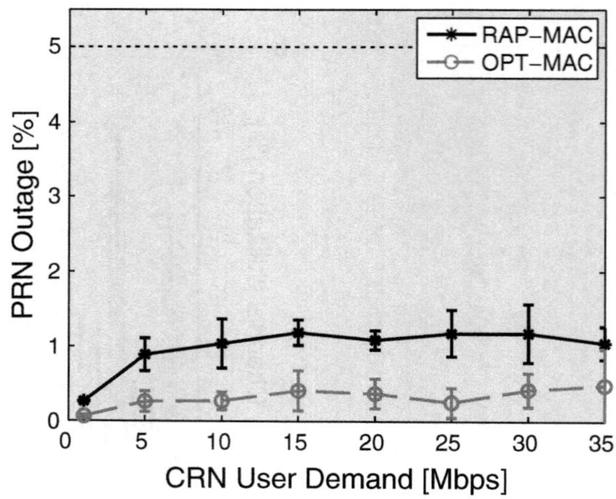

Fig. 7.6 PRN outage probability of PRN 4 for β equals to 5%

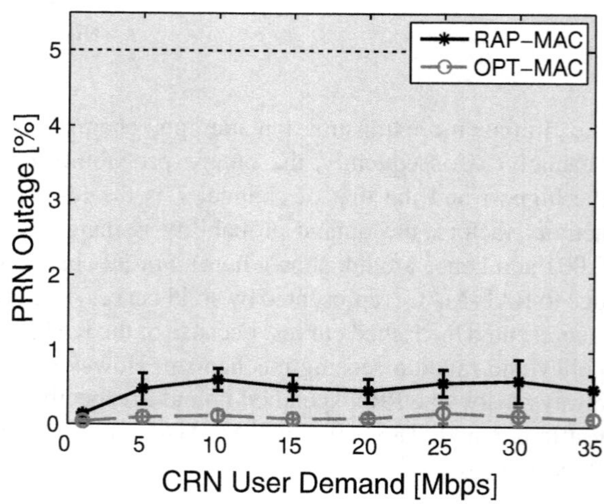

Fig. 7.7 PRN outage probability of PRN 7 for β equals to 5%

References

1. MATLAB: version 7.10.0 (R2010a). The MathWorks Inc., Natick (2010)
2. Salameh, H.B., Krunz, M., Younis, O.: MAC protocol for opportunistic cognitive radio networks with soft guarantees. IEEE Trans. Mobile Comput. **8**(10), 1339–1352 (2009)
3. Huang, S., Liu, X., Ding, Z.: Optimal transmission strategies for dynamic spectrum access in cognitive radio networks. IEEE Trans. Mobile Comput. **8**(12), 1636 –1648 (2009)
4. Zhao, Q., Tong, L., Swami, A., Chen, Y.: Decentralized cognitive MAC for opportunistic spectrum access in ad hoc networks: A POMPD framework. IEEE J. Sel. Area. Comm. **25**(3), 589–600 (2007)
5. Chaporkar, P., Proutiere, A., Asnani, H.: Learning to optimally exploit multi-channel diversity in wireless systems. In: Proceedings of IEEE INFOCOM 2010, San Diego, CA (2010)
6. Jain, R.: The Art of Computer Systems Performance Analysis. Wiley, New York (1991)

Chapter 8
Hardware Implementation

8.1 Hardware Platform

In this section, we specify the requirements of candidate hardware platforms that can be used to implement Opportunistic Spectrum Access protocols such as the proposed RAP scheme and briefly overview the exiting platforms. Then, we explain in more depth the technical details of the platform we chose for our RAP-MAC implementation.

8.1.1 Platform Requirements

Two main features are necessary for a candidate platform to be eligible to implement Opportunistic Spectrum Access protocols:

- *Cognitive capability* that enables the platform to infer the current occupancy of the spectrum. The spectrum utilization information should be continuously available and updated at the spectrum allocation module of the platform so that the appropriate transmission parameters are set. Two spectrum sensing approaches can be used: wide-band sensing and narrow-band sensing. Wide-band sensing requires a multi-gigahertz front-end transceiver to scan the entirety of the spectrum. However, wide-band sensing results in delayed spectrum utilization information that affects the accuracy of the spectrum access decision. On the other hand, narrow-band sensing only investigates the utilization of a small portion of the spectrum, and hence, can lead to missing spectrum access opportunities. However, the fast response of narrow-band sensing better tracks the dynamic nature of spectrum utilization. Our empirical performance evaluation study assumes cognitive radios with narrow-band sensing capability as the case with the wide-range of commercially available transceivers.

A. Khattab et al., *Cognitive Radio Networks: From Theory to Practice*, Analog Circuits and Signal Processing, DOI 10.1007/978-1-4614-4033-8_8,

- *Re-configurability* of the RF module of the platform that allows the operating parameters of the transceiver to be configured on the fly (i.e., in real-time) without making any changes to the hardware components that affect the radio emission. The main transceiver parameters to be configured are the operating frequency band, modulation type and transmission power.

8.1.2 Overview of Existing Platforms

Existing hardware/software platforms that can be used to implement Opportunistic Spectrum Access protocols can be classified into two main classes: Software-Defined Radio (SDR)-based and Field Programmable Gate Array (FPGA)-based platforms. SDR platforms are implemented via the integration of the GNU Radio that is a software development environment [1] and any of the Universal Software Radio Peripheral (USRP) product family that is used as the RF interface of the platform [2]. SDR platforms provide seamless flexibility in implementing spectrum sensing and spectrum management since they rely on software to implement such functionalities. Different open-source GNU radio software specifically written for CRNs are available such as the Cognitive Radio Open Source System (CROSS) [3], and the Papyrus software platform [4]. However, the goodput and latency of the prototypes implemented via SDR platforms are one to three orders of magnitude worse than realistic hardware designs and lag far behind the requirements of real-world communication schemes such as IEEE 802.11 [5, 6]—despite their low cost. This latency is attributed to the use of Universal Serial Bus (USB) cables to interface the low-cost USRP hardware to the general purpose computer that carries out the SDR signal processing. Consequently, such platforms achieve peak rates that are barely close to the state-of-the-art commercial systems. Hence, such approaches lack the performance needs of realistic systems.

On the other hand, FPGA-based platforms offer orders of magnitude improvement in the latency and goodput performance at the expense of increased hardware complexity and cost. An FPGA-based platform is often composed of a hardware component that consists of a compact FPGA board which implements the physical and link layers associated with a software environment that provides the basic physical and MAC layers functionalities and interfaces to the hardware component in order to allow the researchers to program the hardware as desired. Thus, FPGA-based systems combine the programmability of software and the high performance and predictability of hardware. Although other platforms exists (e.g., AirBlue [6]), we have chosen the Wireless open-Access Research Platform (WARP) [7] FPGA-based platform for our RAP-MAC implementation.

8.2 WARP Platform Overview

We use the Wireless open-Access Research Platform (WARP) to implement the proposed RAP-MAC protocol in order to demonstrate its performance gain in practice. WARP is the state-of-the-art of wireless experimentation tools that is well recognized by both the academic and industrial research communities. The WARP physical (PHY) layer is custom designed to fulfill the needs of high-performance wireless communications systems. The WARP platform is designed to facilitate the development and performance evaluation of a large class of cross-layer designed protocols by adopting a flexible interrupt-driven interface between the PHY and the medium access control (MAC) layers. This contrasts with (1) platforms that are based on off-the-shelf wireless interfaces which limit experimentation only to enhancements of existing standards currently implemented, and (2) SDR platforms that are limited to low-speed applications due to hardware/software interfacing bottlenecks.

8.2.1 WARP Hardware Components

For our RAP-MAC implementation we use the WARP FPGA board version 2.2 shown in Fig. 8.1. The main components of this version of the WARP FPGA board that are relevant to our RAP-MAC implementation can be summarized as follows.

Fig. 8.1 WARP FPGA board [7]

- **Xilinx Virtex-4 FPGA:** The WARP board is built around the Xilinx Virtex-4 FPGA which compromises 42 K logic slices, 160 DSP48 slices, 376 block RAMs, 4 Tri-mode Ethernet MACs and 2 PowerPC cores. The physical layer protocol implementations are directly implemented on the FPGA fabric. Meanwhile, the higher layer protocols such as the medium access protocols are written in C-language and downloaded to one of the FPGA PowerPC cores where they directly communicate with the physical layer peripherals implemented using the FGPA fabric. The WARP open-access repository provides the FPGA implementation of an Orthogonal Frequency Division Multiplexing (OFDM) physical layer alongside a set of low-level physical layer control and MAC primitives. These primitives create a flexible interface that provides user-level access to the parameters commonly required for cross-layer designs. Thereby, WARP provides seamless flexibility in the development of Opportunistic Spectrum Access protocols such as RAP-MAC that require cross-layer interaction. For our RAP-MAC implementation we use the WARP OFDM Reference Design version 14.
- **Radio Card:** The WARP FPGA board has four identical daughtercard slots. The WARP hardware supports any combination of daughtercards in the four slots. For our experiments, we only need the WARP radio daughtercard. Up to four radios can be mounted on a single WARP board to enable 4×4 MIMO systems. The WARP radio boards are custom designed to operate over both the 2.4 GHz and 5 GHz ISM bands. The implemented dual-band radio transceiver is intended for OFDM applications with a bandwidth up to 40 MHz. For the WARP reference design used in this work, WARP uses a 64-subcarrier OFDM physical layer supporting BPSK, QPSK, and 16 QAM modulation schemes within a bandwidth of 10 MHz.
- **10/100/1000 Ethernet port:** The WARP FPGA Board is equipped with one 10/100/1000 Ethernet device. This Ethernet port is connected to one of the hardened Tri-mode Ethernet MAC cores on the Virtex-4 FPGA. The WARP Ethernet port serves as the interface to the application layer. Data packets to be transmited/received over the air by WARP are sourced/sinked by an Ethernet enabled device (e.g., a computer or a router) that is connected to a WARP node via the WARP Ethernet port.

8.3 RAP-MAC Implementation

In this section, we describe the design steps of implementing the proposed RAP-MAC protocol using the WARP platform. In order to illustrate the different attributes of the performance gain of RAP-MAC we also implement a suite of Opportunistic Spectrum Access protocols. We first develop a custom cross-layer framework for implementing opportunistic spectrum management mechanisms that combines the basic functionalities required by a wide range of OSA protocols. Then we describe the detailed state machine of our RAP-MAC protocol implementation. Finally, we briefly describe the implementation details of the other opportunistic spectrum management protocols used for comparison.

8.3.1 General OSA Implementation Framework

We first instrument the basic functionalities commonly used by different opportunistic spectrum management schemes. We implement the following four mechanisms using the WARP OFDM reference design version 14: (1) spectrum sensing, (2) common control channel, (3) spectrum coordination packet handshake, and (4) multi-rate multi-power packet transmission.

- **Spectrum Sensing.** The purpose of this function is to measure the cumulative interference of a given spectrum band and determine whether it is below the power mask of the corresponding primary network or not. This is realized by monitoring the received signal strength indicator (RSSI) averaged over a certain time window. By comparing the time-averaged RSSI with the spectrum power mask, an Opportunistic Spectrum Access protocol can determine whether this band is clear $(P_{int}^{(i)} \leq P_{mask}^{(i)})$ or not $(P_{int}^{(i)} > P_{mask}^{(i)})$.
- **Common Control Channel.** Distributed Opportunistic Spectrum Access protocols require a means by which a cognitive sender coordinates its spectrum decisions with its intended receiver. A common control channel is generally used for this purpose. Both the sender and the receiver are continuously listening to this channel if not involved in an active data exchange. We define channel 14 of the 2.4 GHz ISM band as the common control channel. Channel 14 of the 2.4 GHz band is not available for commercial purposes in the United States and can only be used for academic research. Using such a channel guarantees a robust common control channel.
- **Spectrum Coordination Packet Handshake.** We create the control packets to be exchanged over the common control channel for cognitive sender-receiver coordination. These control packets do not include any payload bytes and only include the sender and the intended receiver addresses in addition to other protocol-dependent control information such as the selected spectrum, the measured RSSI, the modulation rate, etc. For RAP-MAC and the other tested OSA protocols, we only need a two-way control-message handshake in which the sender informs its receiver with its spectrum selections via a control packet and the receiver accepts or declines such selections with another control packet. The control packet handshake is transmitted using the base rate realized via the WARP QPSK modulation rate.
- **Multi-rate Multi-power Packet Transmission.** Finally, we implement a data packet transmission scheme which parameters are configured on a packet-per-packet basis. For the considered opportunistic spectrum management schemes we allow the protocol to configure the transmission channel, the modulation rate and power. A data packet can use one out of three WARP modulation schemes: BPSK, QPSK, and 16 QAM with respective transmission powers of 12 dBm, 15 dBm, and 18 dBm. The effective rates of these three modulations schemes were measured to be 4.1 Mbps, 8.4 Mbps, and 10.4 Mbps, respectively.

8.3.2 RAP-MAC State Machine

We implement the RAP-MAC protocol using the aforementioned mechanisms. We write the RAP-MAC sender and receiver state machines, depicted by the flow charts in Figs. 8.2 and 8.3, respectively, in C-language and compile them in a single bit file to be downloaded to one of the WARP FPGA PowerPC cores. A demonstration of the RAP-MAC implementation in action is available online [8].

8.3.2.1 RAP-MAC Sender State Machine

RAP-MAC nodes are continuously listening to the common control channel in case an intended control packet is destined to them. If the application layer of a node has a packet to be transmitted, the RAP-MAC node switches to the sender state machine depicted by the flow chart in Fig. 8.2. A RAP-MAC sender first checks if there exists a favorite spectrum that recently carried out a successful transmission. If such a favorite spectrum exists, RAP-MAC moves to the SR packet transmission state. Otherwise, RAP-MAC randomly picks a new spectrum then moves to the SR packet transmission state. In the SR packet transmission state, RAP-MAC tunes the radio transceiver to the selected spectrum and measures the received signal strength of the selected spectrum over a period of 50 μs. Then, the sender creates a SR packet that includes a single-bit indictor of whether the measured interference is below the power mask of the spectrum or not. The sender then re-tunes the transceiver to the common control channel, sends the SR packet, sets a timeout timer, and enters the idle listening state on the common control channel.

If the intended receiver does not reply back with a SG packet before the timeout timer expires, the RAP-MAC sender increases the number of transmission attempts of the packet before choosing a new spectrum other than the one that was just probed. The RAP-MAC sender then re-enters the SR packet transmission state for the newly selected spectrum.

If the sender receives a SG packet before the expiration of the timeout timer, it switches to the data packet transmission state. In this state, the RAP-MAC sender extracts the receiver's assessment of the spectrum (whether it can be considered as a clear or unclear spectral opportunity) as well as the receiver's suggested transmission rate from the SG packet. In the case of an unclear opportunity, the sender re-enters the SR packet transmission state after selecting a new spectrum with probability $1-q$. Otherwise, the RAP-MAC sender enters the data packet transmission state on the selected spectrum with a rate probabilistically chosen as described in Chap. 5. If an acknowledgment packet is received within a timeout period and the spectral opportunity is a clear one, the RAP-MAC sender declares this spectrum as it favorite for the upcoming transmission. Otherwise, no favorite spectrum is defined. The RAP-MAC then returns to the idle listening state on the common control channel.

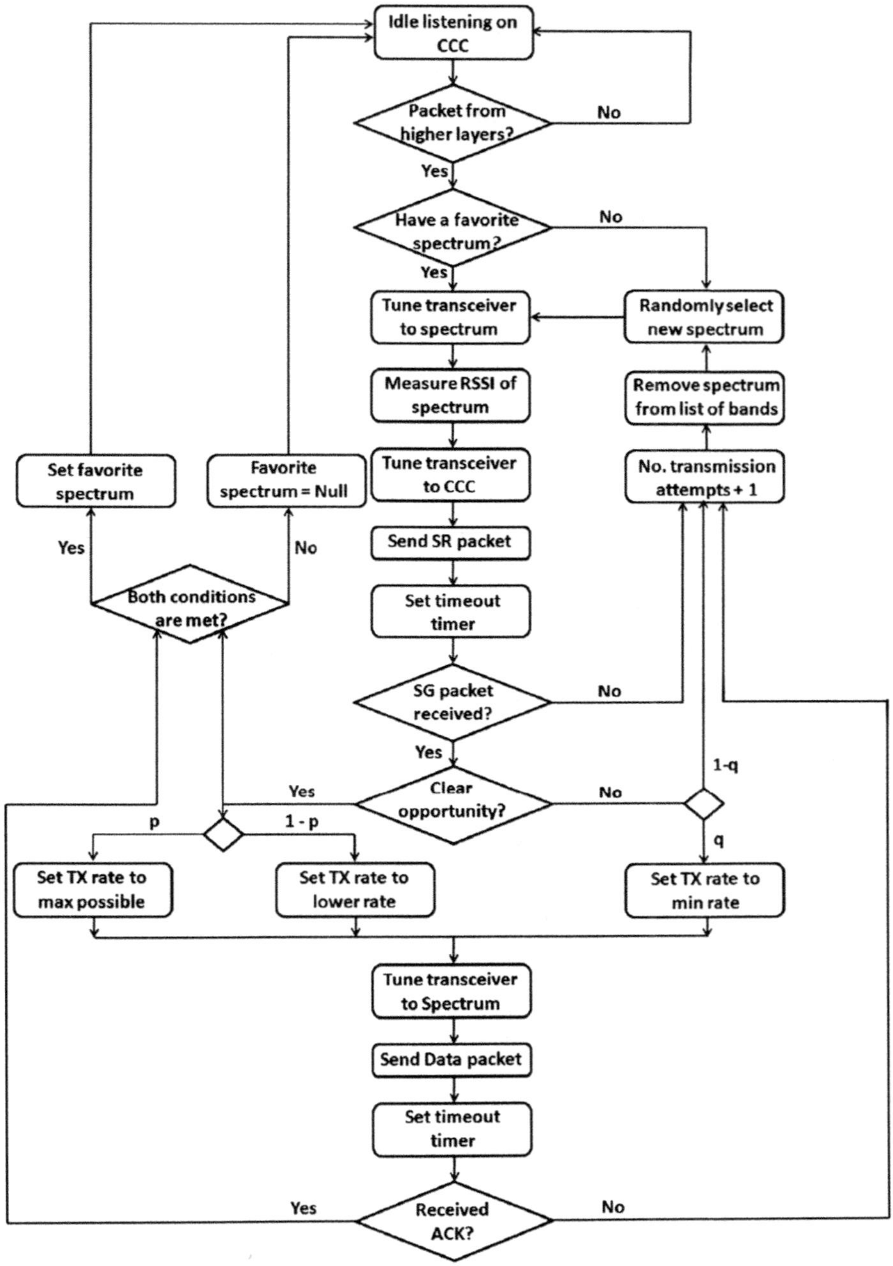

Fig. 8.2 RAP-MAC transmitter state machine

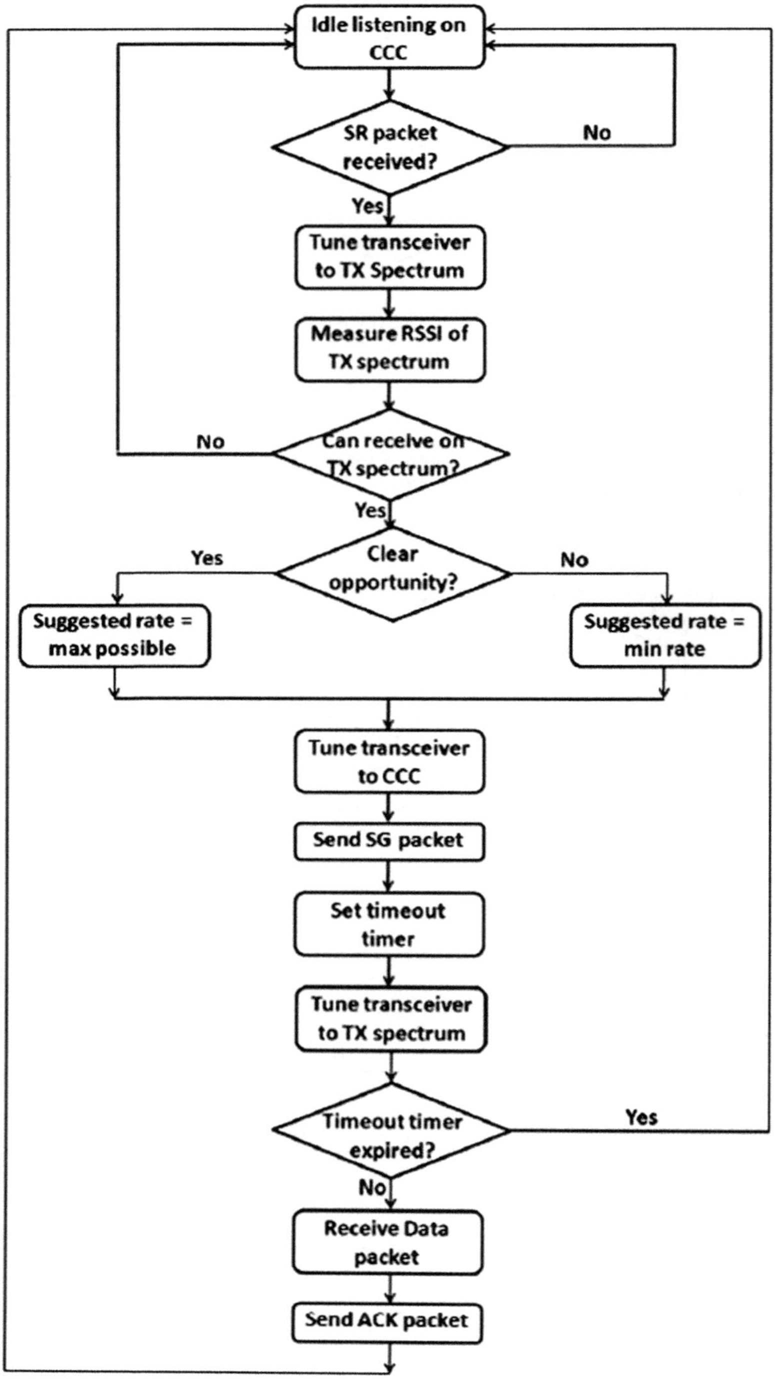

Fig. 8.3 RAP-MAC receiver state machine

8.3.2.2 RAP-MAC Receiver State Machine

Alternatively, a RAP-MAC node can receive a SR packet destined to it while being in the idle listening state on the common control channel. In this case, the node moves to the receiver state machine depicted by the flow chart in Fig. 8.3. After decoding the SR packet, the RAP-MAC node tunes the radio transceiver to the sender-selected spectrum indicated by the SR packet. If the cumulative RSSI at the receiver side is above the power mask of the spectrum, implying a strong interference level on the sender-selected spectrum at the receiver vicinity, the receiver returns to the idle listening state on the common control channel without sending a SG packet. However, if the RSSI measurement indicates a low interference level on the sender-selected spectrum, the receiver enters the SG transmission state. The receiver includes the maximum rate it can sustain given the RSSI measurement as well as an indicator of whether or not this spectral opportunity is clear. The RAP-MAC receiver re-tunes its radio to the common control channel, sends the SG packet, and sets a timeout timer. Then, the RAP-MAC receiver tunes its radio to the selected data channel. If no packet is detected within the timeout period, the RAP-MAC receiver re-enters the idle listening state on the common control channel. This means that the sender probabilistically decided not to transmit on the selected spectrum. On the other hand, if a packet is detected, the receiver disables the timeout timer and finishes the packet reception. If the packet is correctly received, the RAP-MAC receiver sends an acknowledgment packet on the data channel then moves back to the idle listening state on the common control channel.

8.3.3 Benchmark Protocols Implementations

In order to empirically evaluate the performance of the proposed RAP OSA approach and how much gain is attributed to its different mechanisms, we also implement three other opportunistic spectrum management schemes.

- **Sequential Sensing with Greedy Access.** This implementation reflects a wide range of existing Opportunistic Spectrum Access protocols (e.g., [9–11]). In such schemes, a cognitive radio node senses all of the available spectrum bands before deciding which band to use. Like other low-complexity single-radio transceivers, the WARP transceiver can be tuned to only one frequency channel at a time. Therefore, we implement a sequential spectrum sensing mechanism in which a cognitive node goes over the channels of interest and reports back the RSSI on individual channels. Unlike the proposed RAP approach, such schemes adopt deterministic and greedy access mechanisms in which a sender only transmits if there exists a spectrum band which its measured RSSI is below the power mask. Furthermore, they transmit using the highest possible power and rate for all the time. Recall that the OPT-MAC protocol used as the benchmark for RAP-MAC performance evaluation in arbitrary large-scale networks using simulations in Chap. 7 is a candidate of this class of protocols.

- **Sequential Sensing with Probabilistic Access.** The second protocol that we use for comparison is a derivative of the above implementation which still depends on sequentially scanning all of the available spectrum bands before deciding the best one to use. However, this protocol adopts a probabilistic and non-greedy spectrum access approach similar to that developed for the RAP-MAC protocol instead of using deterministic and greedy spectrum access. Such a protocol helps identifying how much gain can be achieved through using a probabilistic access mechanism if adopted by the wide range of existing OSA protocols that rely on greedy access strategies. Furthermore, this protocol implementation allows us to assess how much gain of the proposed RAP OSA approach is due to random sensing since the sensing mechanism is the only difference between RAP-MAC and this protocol implementation.
- **Random Sensing with Greedy Access.** We also implement a variant of the proposed RAP-MAC protocol which uses randomized sensing in conjunction with a greedy spectrum access mechanism. We refer to this protocol implementation as the random sensing with greedy access protocol. The greedy access mechanism of this protocol is the same one used by the sequential sensing with greedy access protocol. Hence, this protocol allows us to quantify the performance gain of randomized narrow-band sensing compared to sequential wide-band sensing. Furthermore, comparing the performance of this protocol implementation against the proposed RAP-MAC illustrates the contribution of the RAP-MAC probabilistic access component in the overall RAP-MAC gain as will be demonstrated by our experiments.

References

1. GNU Radio: http://gnuradio.org/redmine/projects/gnuradio/wiki. Accessed 25 July 2012
2. Ettus Research LLC: http://www.ettus.com/. Accessed 25 July 2012
3. CROSS: Cognitive Radio Open Source System. http://cornet.wireless.vt.edu/trac/wiki/Cross. Accessed 25 July 2012
4. Yang, L., Zhang, Z., Hou, W., Zhao, B.Y., Zheng, H.: Papyrus: A software platform for distributed dynamic spectrum sharing using SDRs. ACM SIGCOMM Comput. Comm. Rev. **41**, 31–37 (2011)
5. Nychis, G., Hottelier, T., Yang, Z., Seshan, S., Steenkiste, P.: Enabling MAC protocol implementations on software-defined radios. In: Proceedings of USENIX symposium on NSDI, Boston, MA (2009)
6. Ng, M.C., Fleming, K.E., Vutukuru, M., Gross, S., Arvind, H.B.: ch8:Airblue: A system for cross-layer wireless protocol development. In: Proceedings of ACM/IEEE ANCS '10, Brooklyn, NY (2010)
7. Rice University WARP Project: http://warp.rice.edu. Accessed 25 July 2012
8. RAP-MAC demo: http://www.youtube.com/watch?v=8X-cbQFrLxM. Accessed 25 July 2012
9. Huang, S., Liu, X., Ding, Z.: Optimal transmission strategies for dynamic spectrum access in cognitive radio networks. IEEE Trans. Mobile Comput. **8**(12), 1636 –1648 (2009)
10. Zhao, Q., Tong, L., Swami, A., Chen, Y.: Decentralized cognitive MAC for opportunistic spectrum access in ad hoc networks: A POMPD framework. IEEE J. Sel. Area. Comm. **25**(3), 589–600 (2007)
11. Chaporkar, P., Proutiere, A., Asnani, H.: Learning to optimally exploit multi-channel diversity in wireless systems. In: Proceedings of IEEE INFOCOM 2010, San Diego, CA (2010)

Chapter 9
RAP-MAC Empirical Performance Evaluation

9.1 Experimental Setup

Implementing a CRN environment poses significant design challenges. For example, any Opportunistic Spectrum Access experiment requires the creation of multiple primary networks which spectral opportunities can be exploited by the CRN users when the primary users are inactive. Thus, the experiments must provide controllable primary network flows. Furthermore, the experimental setup must keep track of every CRN transmission as well as every primary networks' transmission and reception in order to assess the CRN decision mechanism and the outage performance of the primary networks, respectively.

Primary Networks Implementation. For our experiments, we create two primary networks each composed of a single sender and a single receiver. In order to have full control over the performance of the primary networks and to not harm existing licensed networks, we configure the two primary networks to operate over non-overlapping channels in the unlicensed 2.4 GHz ISM band. More specifically, we configure the first PRN to use channel 1 of the 2.4 GHz and the second PRN to use channel 7 of the same band. We use laptops equipped with IEEE 802.11 wireless cards to create the primary networks. The transmission power of each network is set to 18 dBm and the physical layer transmission rate is set to 11 Mbps with the auto-rate feature turned off. We use *iperf* [1] to generate a UDP flow from each primary sender and collect the UDP flow statistics at the corresponding receiver. We measure the backlog UDP capacity of the two primary networks in the absence of any CRN activities to be 6.03 Mbps and 6.15 Mbps, respectively.

Cognitive Radio Network Implementation. We create a cognitive radio node by connecting a laptop (with its wireless interface disabled) to a WARP board via the WARP Ethernet port. By downloading the appropriate bit file of any of the implemented Opportunistic Spectrum Access protocols to a WARP PowerPC, the WARP will act as the wireless air interface of the laptop that runs that particular OSA protocol [2]. We create a fully backlogged cognitive radio transmission between

A. Khattab et al., *Cognitive Radio Networks: From Theory to Practice*,
Analog Circuits and Signal Processing, DOI 10.1007/978-1-4614-4033-8_9,

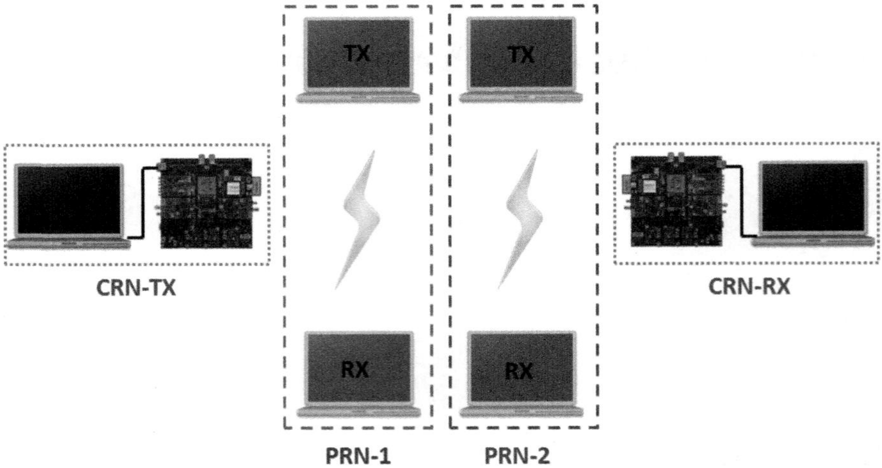

Fig. 9.1 Illustration of the experiment setup

two such cognitive radio nodes using *iperf*. The cognitive radio sender and receiver nodes are at equal distance of approximately 2 m from the senders and receivers of the two collocated primary networks. Figure 9.1 depicts a layout of the experiment setup. As explained earlier, a cognitive node chooses form BPSK, QPSK, and 16-QAM modulation schemes with corresponding powers of 12 dBm, 15 dBm, and 18 dBm, respectively. The individual backlogged rates of these three modulation schemes are 4.1 Mbps, 8.4 Mbps, and 10.4 Mbps, respectively. The control packets are transmitted using QPSK modulation at 18 dBm power on channel 14 of the 2.4 GHz ISM band. Table 9.1 summarizes the experiment parameters.

The reported results in the remainder of this chapter are the average of five runs each of 1 min length. We run the experiments between midnight at the early hours of the morning to minimize the potential uncontrolled transmission activities over the used channels.

9.2 Experimental Results

9.2.1 RAP-MAC Capacity

We begin by characterizing the performance of RAP-MAC in the absence of primary networks' activities. We vary the RAP-MAC probabilities p and q and plot the achieved goodput in Fig. 9.2. As shown in Fig. 9.2, the RAP-MAC goodput increases by almost 2 Mbps as p increases from 0 to 1. Recall that p is the probability of transmitting at the highest rate the receiver can sustain in the absence of primary networks' activities. As p increases, the RAP-MAC goodput tends to approach

Table 9.1 Experiment parameter summary

Parameter	Value
Channel switching time	10 μs
Timeout period	50 μs
Sensing time per channel	30 μs
Transceiver turnaround time	23 μs
Payload packet length	1450 Bytes
Control packet length	24 Bytes
TX power (BPSK, QPSK, 16QAM)	(12,15,18) dB
MAC data rate (BPSK, QPSK, 16QAM)	(4.1, 8.4 10.4) Mbps

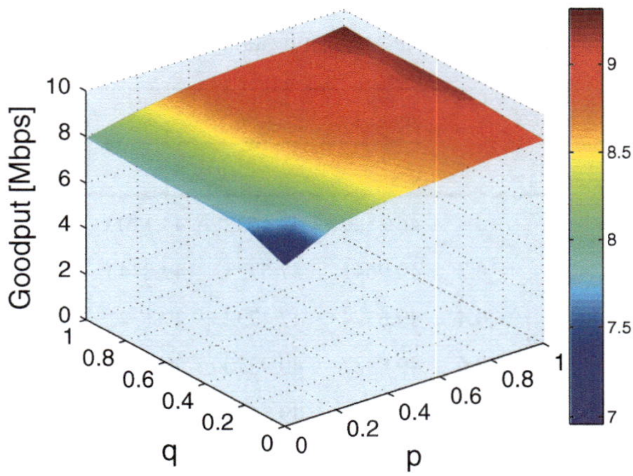

Fig. 9.2 RAP-MAC maximum goodput achieved in the absence of any primary networks' activities

the 16-QAM transmission rate. On the other hand, q has a minimal impact on the RAP-MAC goodput. This is because q is the probability of transmission given the cognitive sender measures primary networks' activities. In this experiment the primary networks are not active at all. Tables 9.2 and 9.3 depict the percentage of the packets transmitted on each channel. The RAP-MAC protocol almost equally uses both channels despite its randomized approach in selecting the data channel. However, it slightly tends to favor channel 1 especially at low q values.

9.2.2 RAP-MAC Parameter Selection

Figure 9.2 implies the use of p and q values of 1 and 1 to maximize the cognitive flow goodput in the absence of the primary networks' activities. However, the scenario considered in the above experiment represents the best-case scenario of inactive primary networks. In this experiment, we consider the worst-case scenario

Table 9.2 RAP-MAC utilization percentage of channel 1 in the absence of PRN activities

		p					
		0	0.2	0.4	0.6	0.8	1
q	**0**	53.6411	52.5358	55.8891	54.9598	57.6606	52.4290
	0.2	59.0359	52.4267	50.1557	49.9651	51.6580	50.9967
	0.4	50.3626	52.5268	50.8279	49.6986	50.0039	50.6888
	0.6	50.1613	50.0730	49.6683	50.2219	50.2665	50.7884
	0.8	50.9419	50.7369	50.4188	49.6944	50.1922	49.7802
	1	50.4121	50.1705	50.1590	50.4721	50.3806	50.4018

Table 9.3 RAP-MAC utilization percentage of channel 7 in the absence of PRN activities

		p					
		0	0.2	0.4	0.6	0.8	1
q	**0**	46.3589	47.4642	44.1109	45.0402	42.3394	47.5710
	0.2	40.9641	47.5733	49.8443	50.0349	48.3420	49.0033
	0.4	49.6374	47.4732	49.1721	50.3014	49.9961	49.3112
	0.6	49.8387	49.9270	50.3317	49.7781	49.7335	49.2116
	0.8	49.0581	49.2631	49.5812	50.3056	49.8078	50.2198
	1	49.5879	49.8295	49.8410	49.5279	49.6194	49.5982

in which both primary networks are fully backlogged and characterize the RAP-MAC performance for different p and q values. Our performance metrics are both the goodput of the CR flow (defined as the amount of data correctly received at the receiver) as well as the outage percentage (defined as the percentage of loss in the transmission rate due to the activity of the cognitive radio transmission) of both PRNs.

Figure 9.3 depicts the goodput achieved by RAP-MAC when both the primary networks' flows are fully backlogged for different p and q values. Unlike the prior experiment wherein p had the dominant impact on the RAP-MAC performance, q has more impact on the RAP-MAC goodput in such a worst-case scenario. Recall that q is the probability of transmission in the presence of primary networks' activities. As q increases from 0 to 0.4, the RAP-MAC goodput increases from 400 to 450 kbps. The RAP-MAC goodput then decreases to 320 kbps as q increases beyond 0.4. This is because the probability of colliding with a primary transmission increases with the cognitive sender attempting more aggressively to exploit such

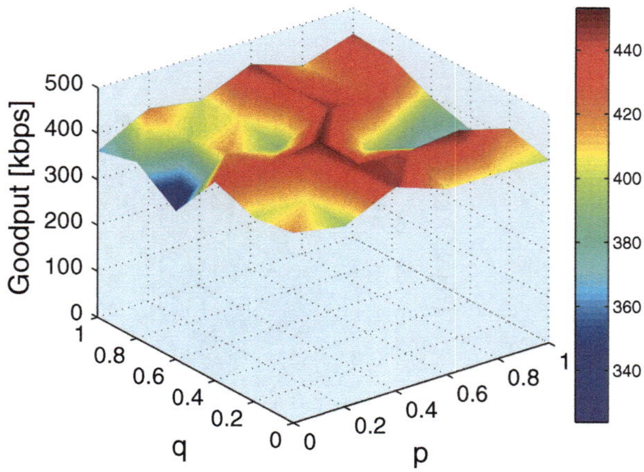

Fig. 9.3 RAP-MAC worst-case goodput achieved when both primary networks are fully backlogged

unclear spectral opportunities. Such a negative impact of increasing the value of q does not only degrade the performance of the cognitive radio transmission but also negatively impact the outage performance of the primary networks as illustrated in Fig. 9.4. As q exceeds 0.4, the primary networks' outages significantly increase. As the value of q approaches unity, the RAP-MAC goodput increases once again as shown in Fig. 9.3. Furthermore, p starts to have a noticeable impact on the RAP-MAC goodput. This is due to the significant primary networks' outages caused by such a high q value which in turns cause the IEEE 802.11 backoff window of both primary networks' flows to increase. These large backoff windows create non-authentic spectral opportunities that are exploited by the cognitive radio flow to obtain such a high goodput.

For the rest of our experiments, we target a 5% maximum primary network outage probability. Based on the results depicted in Figs. 9.3 and 9.4, the p and q values that achieve the highest RAP-MAC goodput while satisfying the primary networks' constraints under the worst-case scenario are 0.4 and 0.4, respectively. We use these values for the rest of our experiments even when the primary networks flows are not-fully backlogged since the resulting PRN outages will always satisfy the targeted constraints given that p and q were computed under worst-case conditions. Recall that RAP-MAC does not interactively adapt its parameter values based on the operating conditions.

Tables 9.4 and 9.5 depict the distribution of the utilization of the used channels. Unlike the above experiment wherein RAP-MAC fairly distributes its transmissions over both channels when both primary networks are idle, RAP-MAC tends to use channel 7 more often compared to channel 1. The percentage of using channel 1 and channel 7 increases from 20 and 70% at low p and q value to 26 and 74% at high p and q. Recall that the primary network using channel 1 suffers more outages compared to the primary network that uses channel 7 as was shown in Fig. 9.4. These outages are due to the interception of the primary transmissions by the cognitive radio transmissions which also causes the cognitive radio transmissions to fail. Consequently, Fig. 9.4 also implies that the cognitive radio transmissions over

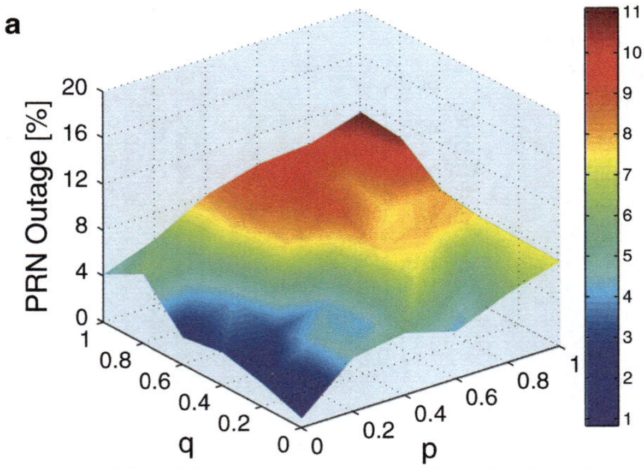

Outage probability of the primary network operating using channel 1.

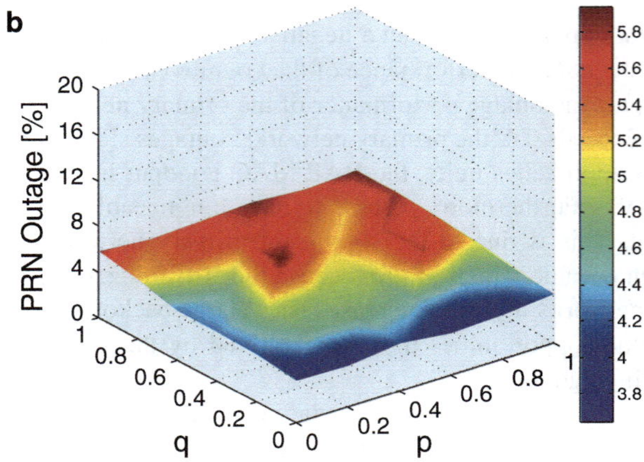

Outage probability of the primary network operating using channel 7.

Fig. 9.4 Worst-case outage probability of the primary networks. For a worst-case outage of 5%, the optimal p and q values are 0.4 and 0.4, respectively

channel 1 are more susceptible to failures compared to the transmissions carried over channel 7. Hence, RAP-MAC tends to favor channel 7 over channel 1 for successful transmissions as shown in Tables 9.4 and 9.5.

9.2.3 RAP-MAC Experimental Performance Evaluation

So far we have characterized the RAP-MAC performance in the best-case and worst-case primary network scenarios to identify its parameters' values satisfying certain PRN outage constraints. Next, we compare the RAP-MAC performance against the other benchmark implementations discussed in Sect. 8.3.3. Our goal is to identify who much each RAP-MAC component contributes to its overall performance gain.

Table 9.4 RAP-MAC utilization percentage of channel 1 with fully backlogged PRN activities

		p					
		0	0.2	0.4	0.6	0.8	1
q	**0**	20.4396	21.2250	23.5232	25.1938	23.3365	26.3240
	0.2	19.0253	21.0404	24.2642	21.7476	24.5469	24.3403
	0.4	23.6269	23.7777	23.5150	27.5574	24.9528	23.7667
	0.6	22.5354	23.5606	25.1991	26.7804	23.4329	24.3036
	0.8	21.0655	22.6349	22.8273	26.3167	26.2525	26.3773
	1	22.4432	23.4443	25.1959	23.4076	23.0478	26.2749

Table 9.5 RAP-MAC utilization percentage of channel 7 with fully backlogged PRN activities

		p					
		0	0.2	0.4	0.6	0.8	1
q	**0**	79.5604	78.7750	76.4768	74.8062	76.6635	73.6760
	0.2	80.9747	78.9596	75.7358	78.2524	75.4531	75.6597
	0.4	76.3731	76.2223	76.4850	72.4426	75.0472	76.2333
	0.6	77.4646	76.4394	74.8009	73.2196	76.5671	75.6964
	0.8	78.9345	77.3651	77.1727	73.6833	73.7475	73.6227
	1	77.5568	76.5557	74.8041	76.5924	76.9522	73.7251

9.2.3.1 CRN Goodput Performance

Figure 9.5 illustrates the googput achieved by the cognitive radio flow according to different protocol implementations. We vary the activity factor of both primary networks by varying the UDP flow rate such that the primary networks' activities go from idle to fully backlogged in 25% activity increments. As shown in Fig. 9.5a, RAP-MAC achieves the highest cognitive flow goodput for all non-zero PRN activities, while the sequential sensing with greedy access approach widely used for Opportunistic Spectrum Access results in the lowest goodput. The RAP-MAC goodput gain increases from 66% at low primary networks' activities to 95% at 50% PRN activity as shown in Fig. 9.5b. As the PRN activities further increase, the RAP-MAC goodput becomes multiple folds of the goodput achieved by the benchmark protocol until the RAP-MAC goodput is 6.7 times the benchmark goodput when the PRNs are fully backlogged.

The superior goodput performance of RAP-MAC is attributed to both its main components: the randomized sensing (which measures the interference on only one

a

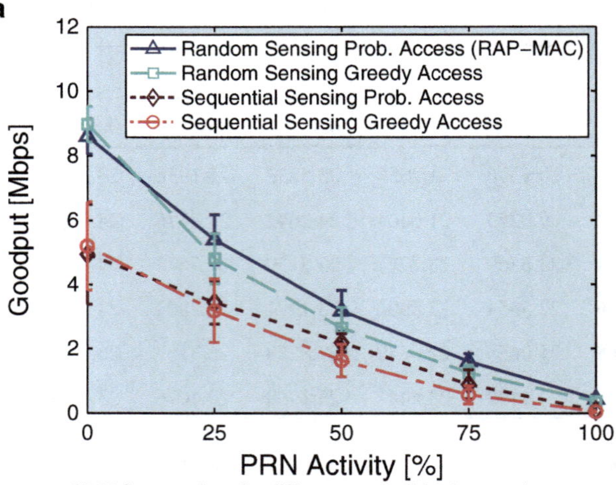

CRN flow goodput for different protocol implementations versus the PRN activity factor.

b

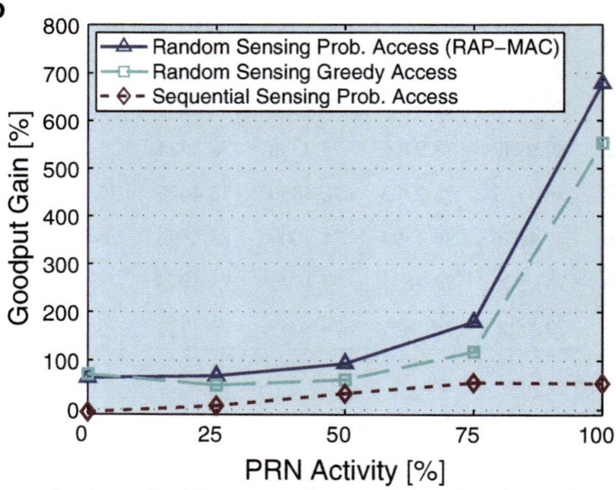

Goodput gain with respect to the sequential sensing with greedy access protocol implementation.

Fig. 9.5 RAP-MAC achieves significant goodput gain over traditional Opportunistic Spectrum Access schemes. While both components contribute to the overall gain, the goodput gain due to randomized sensing is higher than the gain due the probabilistic access mechanism

randomly selected frequency channel instead of scanning all frequency bands) and the non-greedy probabilistic access component (which probabilistically explores the spectral opportunities rather than accessing the channel based on a black or white view). We use the goodput achieved by the other two benchmark protocols to identify how much each component is contributing to the overall gain by performing pairwise comparisons of the achieved goodput shown in Fig. 9.5a.

Intuitively, adopting a greedy access strategy results in a slightly higher goodput, regardless of the sensing mechanism, when the primary networks are idle. Hence, random sensing is the main contributor to the overall gain at low PRN activities as shown by comparing the protocols implementing random sensing against sequential sensing for both access mechanisms. More specifically, by comparing the solid curve (representing RAP-MAC) against the dotted curve (representing its sequential sensing counterpart), and comparing the dashed curve (representing random sensing with greedy access) against the dash-dotted curve (representing the common greedy access based on sequential sensing). As the activities of the primary networks increase, the gain due to probabilistic access increases. For primary networks' activities of 50% and above, the contribution of random sensing is approximately 70–80% of the overall RAP-MAC gain while the contribution of the probabilistic access mechanism is around 20–30% depending on the PRN activities.

Furthermore, the individual components of RAP-MAC can be used to improve the performance of the common spectrum management approaches that use sequential sensing with greedy access as shown in Fig. 9.5b. For example, adopting a non-greedy access mechanism can result in goodput gain of up to 56% as depicted by the dotted curve. Furthermore, exploiting random sensing instead of sequentially searching for the best channel to use achieves 64–82% of the RAP-MAC gain, depending on the PRN activities, as illustrated by the dashed curve. This emphasizes that the random sensing component has a more significant performance gain.

9.2.3.2 PRN Outage Performance

Next, we evaluate the outage performance of the primary networks for different Opportunistic Spectrum Access protocol implementations. Figure 9.6 depicts the outage percentages encountered by both primary networks for all protocol implementations versus the activity of the primary networks. All protocols result in more outages to the primary network using channel 1 as the case with RAP-MAC discussed in Sect. 9.2.2. However, two observations can be made on Fig. 9.6. First, probabilistic access schemes result in only slightly higher PRN outages compared to their greedy access counterparts. However, probabilistic access has a weaker impact on the PRN outage when sequential sensing is used (as illustrated by the small gap between the dotted and dash-dotted outage curves). With the inaccuracies of random sensing, the impact of probabilistic access increases (as illustrated by the gap between the solid and dashed outage curves). Second, random sensing results in approximately 2.6 times the outages due to sequential sensing protocols irrespective of the access protocol. This is because sequential sensing protocols assess the interference levels on both channels before deciding the transmission action. On the other hand, random sensing protocols simply pick a channel at random for transmission. It is worth mentioning that despite resulting in higher primary network outages, random sensing protocols including RAP-MAC adhere to the 5% maximum outage constraint. However, the significant multi-fold goodput gains of such protocol illustrated in Fig. 9.5 outweighs the excess primary outages

a

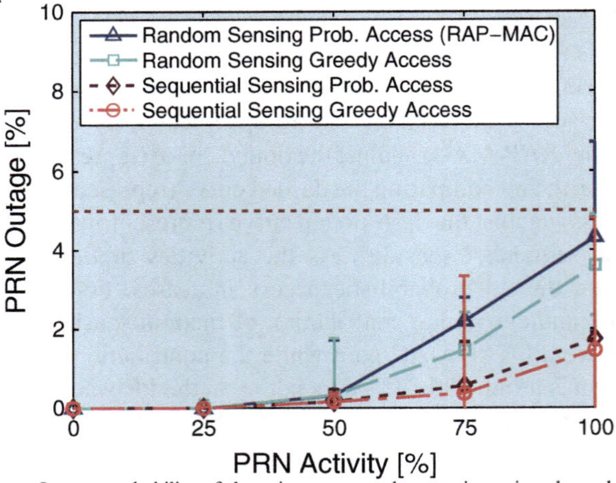

Outage probability of the primary network operating using channel 1.

b

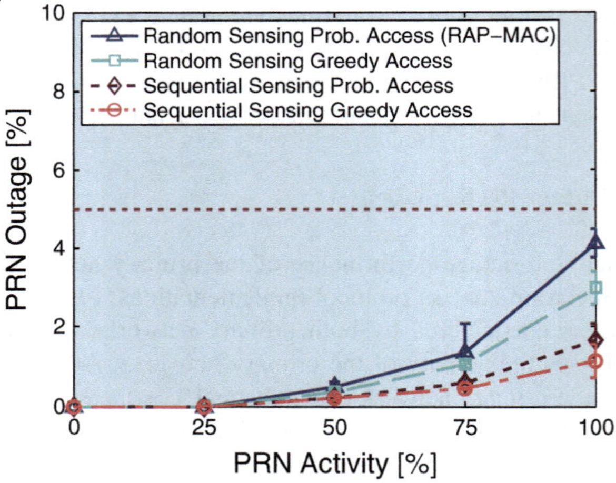

Outage probability of the primary network operating using channel 7.

Fig. 9.6 The outage performance of the primary networks versus the activity factor for different protocol implementations. While both satisfy the 5% PRN outage constraint, random sensing results in more primary outages compared to sequential sensing

resulting from such protocols. Furthermore, as the number of the primary networks increases, the sensing time required to assess the interference on all of the channels will increase. Hence, the RAP goodput gain is expected to further increase.

9.2.3.3 Channel Utilization Pattern

Finally, We investigate the way different Opportunistic Spectrum Access protocols utilize the two available channels. Figures 9.7, 9.8, 9.9, and 9.10 depict the

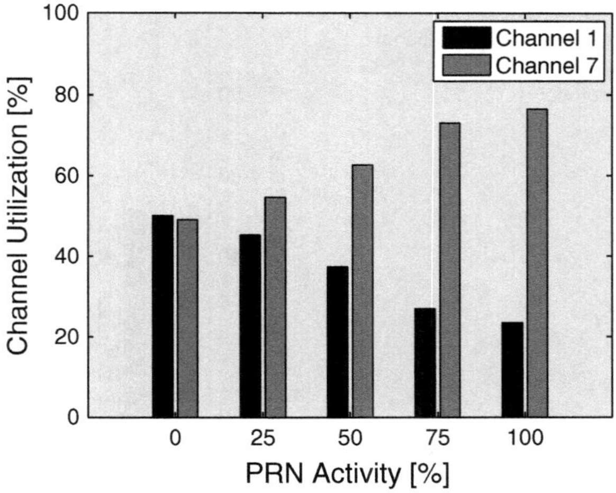

Fig. 9.7 Channel utilization percentage of the RAP-MAC (random sensing with probabilistic access) protocol

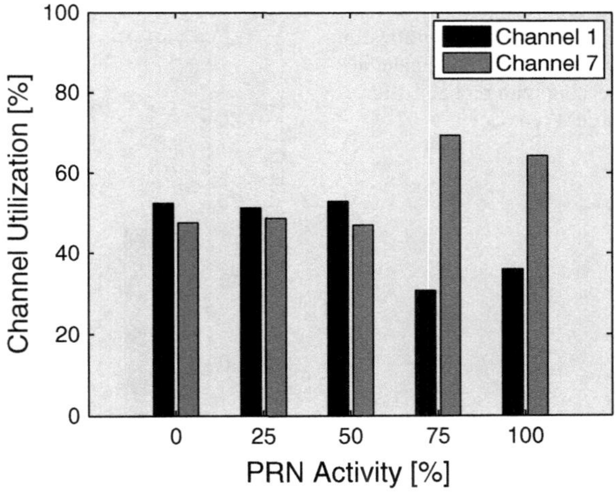

Fig. 9.8 Channel utilization percentage of the sequential sensing with greedy access protocol

utilization percentage of both channels for the RAP-MAC, sequential sensing with greedy access, random sensing with greedy access, and sequential sensing with probabilistic access protocols, respectively. As the activities of the primary networks increase, RAP-MAC tends to have a higher percentage of its packets transmitted over channel 7 (except when the PRNs are idle in which case both channels are evenly utilized) as shown in Fig. 9.7. This behavior confirms our explanation of the channel utilization in the best-case and worst-case scenarios discussed earlier in this chapter. In contrast, the sequential sensing with greedy access strategy commonly adopted by opportunistic access schemes fairly distributes transmissions between the two channels up to 50% PRN activity as depicted in Fig. 9.8. Recall that this protocol results in nearly zero PRN outage probability as long as the activity factor is below 50% as shown in Fig. 9.6. As the activities of the primary network further increase, such a protocol tends to utilize channel 7 more often as the case with RAP-MAC. However, the gap between using channel 7 and channel 1 is smaller than that of the RAP-MAC protocol.

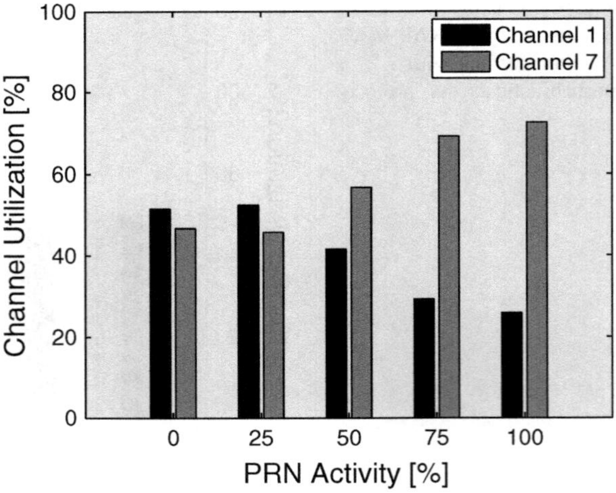

Fig. 9.9 Channel utilization percentage of the random sensing with greedy access protocol

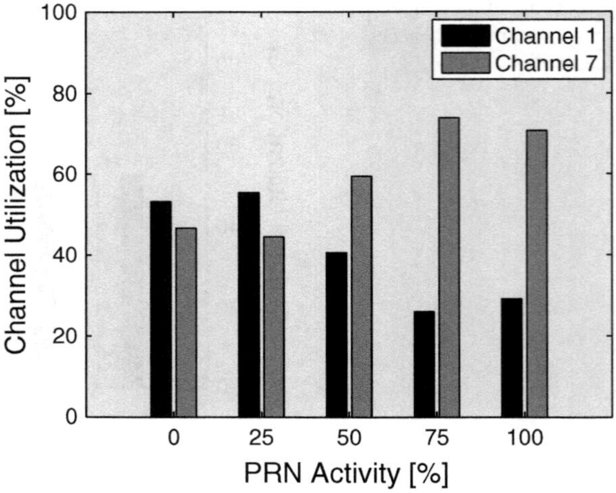

Fig. 9.10 Channel utilization percentage of the sequential sensing with probabilistic access protocol

Meanwhile, Fig. 9.9 shows that substituting the RAP-MAC probabilistic access mechanism with a greedy access one results in smaller gap between the utilization of both channels. This reduction in the gap size is attributed to the reduction in the outage percentage due to the deterministic greedy access approach compared to probabilistic access as shown in Fig. 9.6. Furthermore, greedy access results in more successful transmissions over channel 1 at low activities of the primary networks.

On the other hand, sequential sensing combined with probabilistic access (shown in Fig. 9.10) further increases the gap between the utilization percentage of the two channels compared to that resulting from the traditional sequential sensing with greedy access approach (illustrated by Fig. 9.8). This increase in the utilization percentage gap is due to the higher primary networks' outages caused by the probabilistic access mechanism compared to greedy deterministic access as was shown in Fig. 9.6. Please note that the sequential sensing with probabilistic access protocol also tends to utilize channel 1 more than channel 7 at low primary networks' activities.

References

1. Iperf: http://iperf.sourceforge.net/. Accessed July 25 2012
2. Rice University WARP Project: http://warp.rice.edu. Accessed July 25 2012

Chapter 10
Conclusions and Future Directions

10.1 Conclusions

Opportunistic Spectrum Access in distributed CRNs has received significant research interest, especially from a theoretical perspective. However, the resulting theoretical approaches are challenged by the practical limitations of today's cognitive radios. The focus of this book has been on the less well-studied issue of implementing distributed Opportunistic Spectrum Access techniques given practical radio transceiver technologies. In order to develop a practical and easy-to-implement Opportunistic Spectrum Access approach, we have taken the practical limitation of distributed CRNs into account while formulating the performance optimization problem. Then, we have used a hardware platform that represents the state-of-the-art of wireless transceivers to implement the proposed scheme and demonstrate its superior performance. The contributions and the conclusions of the research presented in this book can be summarized as follows.

- We have presented a framework and a protocol for opportunistic spectrum management. Unlike prior work, we have adopted a probabilistic and non-greedy approach to counter the limitations of distributed CRNs such as the inability to base the spectrum management decisions on the interference at primary receivers and the increased complexity of global network-wide coordination. The proposed Rate-Adaptive Probabilistic (RAP) OSA approach randomly explores spectrum bands. Then it uses a non-greedy probabilistic power/rate selection approach to not only address the inaccuracy in determining the actual interference at nearby primary receivers but also to allow multiple flows to fairly share the spectral opportunities without relying on explicit inter-flow coordination. The randomized spectrum selection technique and the probabilistic spectrum access decisions relax the requirements of the cognitive radio hardware.
- We have analytically formulated the constrained Cognitive Radio Network goodput as a mixed integer non-linear programming problem to find the optimal parameter values that maximize the achievable goodput while satisfy the outage constraints of the primary networks. Solving this mixed integer non-linear

A. Khattab et al., *Cognitive Radio Networks: From Theory to Practice*,
Analog Circuits and Signal Processing, DOI 10.1007/978-1-4614-4033-8_10,

problem for any given values of the system parameters results in the optimal probability of transmission in both the clear and unclear spectrum scenarios as well as the maximum transmission rate and power for each of the used spectrum bands. Our solution implies that the weaker the constraints imposed by the primary networks, the more aggressive the proposed RAP approach will be in exploring the randomly selected spectrum bands. Meanwhile, the RAP approach tends to choose conservative parameter values as the uncertainty of knowing the distribution of the distance between the cognitive radio senders and the primary receivers increases. However, the achievable RAP goodput does not depend on such uncertainty for a given set of constraints imposed by the primary networks.

- With the optimized parameter values, we have developed a packet-level simulator to illustrate the performance of the proposed RAP Opportunistic Spectrum Access approach in arbitrary large-scale networks. Our simulation results have shown that the proposed RAP approach achieves up to 138% improvement in the goodput performance of the cognitive radio users while satisfying the primary network outage constraints in distributed networks compared to greedy spectrum management schemes. This superior goodput performance is attributed to the rate-adaptive access mechanism. Furthermore, RAP-MAC has an outstanding fairness performance, without using an explicit inter-flow coordination, due to its non-greedy transmission policy. On the other hand, the random spectrum selection and the probabilistic access techniques result in more outages (but within the permissible bounds) to the primary networks compared to traditional OSA schemes. However, the superior goodput gain of the RAP techniques outweighs the excess outages of the primary networks.
- Finally, we have demonstrated the superior performance of the proposed practical opportunistic spectrum management in a real system. We have developed an implementation framework of opportunistic spectrum schemes using the Wireless open-Access Research Platform. Using such a framework, we have implemented the proposed RAP spectrum management approach as well as suite of other OSA approaches. We have performed an extensive set of experiments that demonstrated the superior goodput gain (up to multiple folds) of the RAP approach. Using the set of benchmark OSA approaches, we have identified that 70–80% of this gain is due to the RAP random spectrum selection and the remaining 20–30% is attributed to the non-greedy probabilistic access policy. Our experiments have also shown that the excess primary network outages are mainly caused by the random spectrum selection. Moreover, we have demonstrated that existing theoretical OSA approaches can benefit from the performance gains of the individual components of the proposed RAP approach.

10.2 Future Directions

In this book, we have presented the first steps towards realizing practical distributed CRNs based on low-complexity radio transceivers. However, the performance of the proposed Opportunistic Spectrum Access approach can be further improved. In what follows, we discuss two potential enhancement techniques.

10.2.1 *Online Parameter Value Optimization*

One way to improve the performance of the proposed RAP Opportunistic Spectrum Access approach is to compute its optimal parameter values in real-time upon changes in the system parameters. Recall that, the optimal RAP parameter values were computed off-line for a given set of system parameters (e.g., the number of primary networks, the primary and secondary user densities and distributions, the primary network outage constraints, the traffic matrix, etc.) by solving the optimization problem given in (6.21). If any of such system parameters changes (e.g., new primary networks came into play or got disconnected, new outage constraints were imposed, or new traffic patterns are noticed), the values of the RAP-MAC parameters need to be recomputed. This requires the development of distributed algorithms that perform the following two tasks. First, there is need for a distributed scheme to track the changes in the system parameters and trigger the need for new parameter values (e.g. through a set of thresholds). Second, a heuristic solution of this problem is required since the complexity of the RAP-MAC optimization problem given in (6.21) is NP-complete. Furthermore, a distributed implementation of such a heuristic approach is required such that every time a system parameter significantly changes the CRN nodes collaboratively find the new optimal parameter values. The aforementioned distributed schemes will required global network-wide cooperation between the cognitive nodes which further complicates the CRN. For instance, a mechanism through which nodes share their collected system information and/or their computation of the new parameter values is needed. Using the common control channel for such a purpose will deteriorate the CRN network performance and will make the CRN vulnerable to failures and malicious attacks.

10.2.2 *History-Based Randomized Spectrum Selection*

Another potential enhancement of the proposed rate-adaptive probabilistic OSA framework is to track the history of the spectrum usage to have a more informed spectrum selection and minimize the frequency of exchanging the spectrum coordination packets. Recall that, the proposed RAP-MAC protocol has the cognitive sender-receiver pair exchanging the spectrum request and spectrum grant packets prior to every data packet transmission. In order to reduce the associated overhead, and hence, increase the CRN goodput, the cognitive radio nodes can keep a record of the history of prior transmission attempts over the different spectrum bands. Such a record can be used to decide (1) which spectrum bands to randomly choose from, and hence, further reduce the RAP-MAC performance degradation due to the incorrect random decisions, and/or (2) when the spectrum request and grant packets are to be exchanged. The proposed RAP-MAC protocol only has an indicator of whether or not the last used spectrum was able to carry a successful data transmission at a high data rate. If such a successful transmission took place, RAP-MAC is to use the same spectrum for the upcoming transmission. Otherwise, it will randomly pick a new band to probe. Either way, the RAP-MAC sender still checks with its respective receiver if it still can receive data on the selected band. Alternatively, having both the sender and the receiver locally observing the recent history of utilizing different spectrum bands will omit the need for exchanging the spectrum coordination packet on a per-packet basis. For instance, if a certain spectrum is observed to have long periods of availability, it can be used by the cognitive sender-receiver pair for multiple consecutive packet transmissions after a single spectrum request/grant packet exchange. Furthermore, when a new spectrum is to be probed, such a history record will narrow down the number of bands to randomly choose from to only those that are highly likely to lead to a successful transmission. Hence, it will further improve the goodput performance of the RAP approach. It is worth mentioning that such a history-based random spectrum selection does not directly aim at accurately tracking the behavior of the primary network users as the case with [1]. However, it only serves as a guide for the RAP randomized spectrum selection component to increase its hit ratio and improve the overall RAP performance.

Reference

1. Zhao, Q., Tong, L., Swami, A., Chen, Y.: Decentralized cognitive MAC for opportunistic spectrum access in ad hoc networks: A POMPD framework. IEEE J. Sel. Area. Comm. **25**(3), 589–600 (2007)

Glossary

Clear Spectral Opportunity A frequency band which the local interference measurements at both the cognitive radio sender and receiver indicate total power below the power mask defined by the primary licensed owner of that band.

Licensed Spectrum A frequency band that is restricted for exclusive use rights for a single entity through a spectrum licensing authority.

Missed Spectral Opportunity A frequency band that is mistakenly thought to be currently in use by its primary licensed owner due to the measurement of high transmission power over this band. However, using such a band by the Cognitive Radio Network would have not only did not cause harm to the primary network transmission but also resulted in successful cognitive radio transmission.

Miss-predicted Spectral Opportunity A frequency band that is mistakenly thought to be currently not used by its primary licensed owner due to the lack of transmission power over this band. However, using such a band by the Cognitive Radio Network results in interference with the hidden transmission activity of the primary network.

Spectral Opportunity A term that is used to describe a frequency band that is not currently used by the licensed owner of that band.

Unclear Spectral Opportunity A frequency band which the local interference measurements at the cognitive radio sender indicates total power above the power mask defined by the primary licensed owner of that band, meanwhile, the interference measurement at the cognitive radio receiver indicates total power below the power mask.

Unlicensed Spectrum A frequency band that is not restricted for exclusive use rights to a single entity through a spectrum licensing authority. However, the spectrum licensing authority defines the set of rules for both the hardware and deployment methods in such a manner that allows multiple wireless services to simultaneously use such a frequency band without interfering with each other.

A. Khattab et al., *Cognitive Radio Networks: From Theory to Practice*,
Analog Circuits and Signal Processing, DOI 10.1007/978-1-4614-4033-8,

About the Authors

Ahmed Khattab received his B.Sc. (honors) and M.Sc. in Electronics and Communications Engineering from Cairo University, Egypt, in 2002 and 2004, respectively. He received the Master of Electrical Engineering degree from Rice University, and his Ph.D. in Computer Engineering from the University of Louisiana at Lafayette, USA, in 2009 and 2011, respectively. He is currently an Assistant Professor at the Electronics and Electrical Communications Engineering Department in Cairo University. He was a research associate at CACS. His research interests are in wireless communication networks and embedded systems. Specific areas include medium access control, cross-layer PHY/MAC designs, congestion control, resource management, quality of service provisioning, performance evaluation and prototyping of wireless communication systems with emphasize on the IEEE 802.11 multi-antenna and multi-hop networks and Cognitive Radio Networks. He has multiple coauthored and single-authored books, patents, and journal and peer-reviewed conference publications. He served as a technical reviewer for various journals and conferences, including the IEEE/ACM Transactions on Networking, IEEE/ACM Transactions on Mobile Computing, IEEE Transactions on Wireless Communications, IEEE INFOCOM 2006–2009,

A. Khattab et al., *Cognitive Radio Networks: From Theory to Practice*,
Analog Circuits and Signal Processing, DOI 10.1007/978-1-4614-4033-8,

ACM MobiCom 2006–2009, ACM Sigmetrics 2007, WiOpt 2007, and IEEE VTC 2012. He won the best student paper award from the IEEE Computer Society—University of Louisiana at Lafayette student chapter in 2010 and in 2011, and was a finalist in the best paper award contest in the IEEE ICCCN 2008 conference.

Dmitri Perkins received his B.S. degree in Computer Science in 1995 from the Tuskegee University and his M.S. and Ph.D. degrees in Computer Science from Michigan State University in 1997 and 2002, respectively. He is currently an Associate Professor at The Center for Advanced Computer Studies (CACS), University of Louisiana at Lafayette. His primary research interests include performance analysis, scalable protocol design, resource allocation, and self-management issues in multi-hop wireless broadband networks, sensor ad-hoc networks and multi-hop Cognitive Radio Networks. Dr. Perkins received the NSF CAREER Award in 2005 and was awarded the Hardy Edmiston Endowed Professorship in 2008.

Magdy A. Bayoumi received the B.Sc. and M.Sc. degrees in electrical engineering from Cairo University, Egypt, the M.Sc. degree in computer engineering from Washington University, St. Louis, and the Ph.D. degree in electrical engineering from the University of Windsor, Ontario. He is the director of the Center for Advanced Computer Studies (CACS) and was the department head of the Computer Science Department, University of Louisiana, Lafayette. He is also the Z.L. Loflin Eminent Scholar Endowed Chair at the Center for Advanced Computer Studies, University of Louisiana, Lafayette, where he has been a faculty member since 1985. He is a fellow of the IEEE. He was the Vice President for Conferences of the IEEE Circuits and Systems (CAS) Society. He was the vice president for the technical activities of the IEEE Circuits and Systems Society and the chairman of the Technical Committee on Circuits and Systems for Communication and the Technical Committee on Signal Processing Design and Implementation. He was a founding member of the VLSI Systems and Applications Technical Committee and was its chairman. He is a member of the Neural Network and the Multimedia Technology Technical Committees. He was an associate editor of Circuits and Devices Magazine, the IEEE Transactions on Very Large Scale Integration (VLSI) Systems, the IEEE Transactions on Neural Networks, the IEEE Transactions on Image Processing, and the IEEE Transaction on Circuit and Systems II: Analog and Digital Signal Processing and Integration. Dr. Bayoumi served on the Distinguished Visitors Program for the IEEE Computer Society, 1991–1994 and the Circuits and Systems, 1999–2001. Dr. Bayoumi is the recipient of the 2009 IEEE Circuits and Systems Meritorious Service Award and the IEEE Circuits and Systems Society 2003 Education Award. He won the Researcher of the year award and the Distinguished Professor award from the University of Louisiana at Lafayette in 1988 and 1993, respectively. Dr. Bayoumi served on the technology panel and advisory board of the US Department of Education project, "Special Education Beyond 2010," 1990–1993. He was the vice-president of the Acadiana Technology Council. He was on the governor's commission for developing a comprehensive energy policy for the State of Louisiana. He represented the CAS Society on the IEEE National Committee on Engineering R&D policy, IEEE National Committee on Communication and Information Policy, and IEEE National Committee on Energy Policy. Dr. Bayoumi research interests include VLSI design and architectures, low power circuits and systems, digital signal processing, neural networks, and wireless ad-hoc and sensor networks.

Index

A. Khattab et al., *Cognitive Radio Networks: From Theory to Practice*, Analog Circuits and Signal Processing, DOI 10.1007/978-1-4614-4033-8,

Batch number: 10614746

Printed by Printforce, the Netherlands